For the People's Pleasure

AUSTRALIA'S BOTANIC GARDENS

To my Dad whom I love very much
for your 58th Birthday.
Mike
xxx

Note:-My section is shown on page 160, areas 9,10 and 17 also shown
on page 164. (beds at the back of the photos.)

PRINCIPAL
PHOTOGRAPHER
Ron Berg

The Oak Lawn, Royal Botanic Gardens, Melbourne. *(Ron Berg)*

For the People's Pleasure

AUSTRALIA'S BOTANIC GARDENS

Carol Henty

GREENHOUSE PUBLICATIONS

PHOTOGRAPHY FOR WOLLONGONG, BALLARAT & GEELONG
Pamla Toler

PHOTOGRAPHY FOR DARWIN
Terry Knight

COVER PHOTOGRAPH
Ron Berg

COVER & BOOK DESIGN
Lynn Twelftree

First published in 1988 by
Greenhouse Publications Pty Ltd
385-387 Bridge Road
Richmond Victoria 3121 Australia

© Carol Henty, 1988

Maps by Lynn Twelftree
Typeset in 10½/12 pt Palatino by
Meredith Typesetters, Melbourne
Printed in Hong Kong by
South Seas International Press Ltd

National Library of Australia
Cataloguing-in-Publication data:

Henty, Carol.
    For the people's pleasure: Australia's botanic gardens.

    Bibliography.
    Includes index.
    ISBN 0 86436 131 9.

    1. Botanical gardens — Australia. I. Title.

580'.74'494

Photograph on opening page: Eucalyptus caesia, a Western
Australian native. *(Ron Berg)*

Photograph on contents page: Crotalaria cunninghamii, the
green bird flower, in the Kimberley display glasshouse.
(Kings Park and Botanical Gardens)

CONTENTS

ACKNOWLEDGEMENTS

The idea for this book arose from a discussion between Patricia Rolfe and me, and took shape through subsequent discussions. For help in the preparation of the book, my thanks go to Gay Stanton for her horticultural advice at many times; to Elizabeth Douglas; to Margaret Barrett for her initial editing of the book; to Marg Bowman, who completed the editing and saw it through to the end; to Lynn Twelftree for the design; and to Sally Milner, who gave much encouragement.

It would not have been possible to produce this book without the help and cooperation of the directors and curators of the gardens included. For their patience, their generous gift of historical material and their efforts in reading and correcting the chapters on their particular gardens I offer my thanks.

At the Botanic Gardens, Sydney my thanks go to Professor Carrick Chambers and Dr Lawrie Johnson for their help, and my particular thanks to Don Blaxell, who read and corrected the manuscript. I am grateful for the help of botanists Joy Thompson, Tony Rodd and Dr Surrey Jacobs, for that of Ed Wilson, Tony Curry and of the staff at the identification desk. My particular thanks to David Churches of the Public Works Department of the New South Wales Government for his generous help with historical material and for checking information, and to Ingrid Mather, also of the Public Works Department. I would also like to thank Lionel Gilbert for providing historical information and helping with corrections, and Joanna Capon for information on Parramatta.

I extend my sincere thanks to the following: in Wollongong, Deane Miller; in Hobart, Anthony May, and Marcus Hurburgh, who supplied historical information; in Melbourne, Dr David Churchill, and Dr Peter Lumley, Roger Spencer and John Taylor; in Geelong, Ian Rogers; and in Ballarat, Robert Whitehead; in Brisbane, Harold Caulfield and Ross McKinnon, who generously supplied much historical material, and Dr Bob Johnson of the Queensland Herbarium; in Rockhampton, Tom Wyatt, and Tom and Molly Bencke for their historical help; in Townsville, I am grateful to Ian Thomas, Jim Lai and Dorothy Gibson-Wilde, historian; in Cairns, Vince Winkel, Robert Guthrie, and the present director of the Botanic Gardens, James Malcolm; in Adelaide, Dr Brian Morley, and Thekla Reichstein particularly, for her unfailing help; in Darwin, George Brown; in Western Australia, Dr Paul Wycherley, who gave me his knowledge and help, and at the Western Australian Herbarium, Dr John Green; and in Canberra, Estelle Canning of the Herbarium, Arthur Court and Dr Robert Boden particularly, for his help with the project, also Dr Lindsay Pryor, David Shoobridge and John Wrigley.

I would like to thank the following for their horticultural and editorial help: John Patrick particularly, Howard Tanner, his wife Mary, Loraine Browne, Brian Adams, and Professor William Stearn.

I am grateful for access to special sections of the state public libraries for historical information, including the W. L. Crowther Library, the State Library of Tasmania; the J.S. Battye Library of Western Australia; the State Library, Darwin; and the La Trobe Collection, the State Library of Victoria; I thank also the staff at the reference section of the State Library of New South Wales; the library of the Royal Botanic Gardens, Sydney; and Susan Tompkins of Bibliophile, Sydney; the National Library of Australia and its staff members; and the Royal Botanic Gardens, Kew.

For permission to reproduce photographs, I thank the following organisations: the Royal Botanic Gardens, Kew; the Fielding-Druce Herbarium, Oxford; the National Library of Australia; the Museum of Applied Arts and Sciences, Sydney; and the W.L. Crowther Library of the State Library of Tasmania.

In travelling to the gardens I became increasingly aware of the talent of the men and women who made the gardens and this book is a tribute to them.

INTRODUCTION

After the establishment of a government house, the next public amenity to be provided in a new British colony in Australia was usually a garden. Frequently built on land adjoining the governor's residence, it enlarged his 'demesne' visually, and made a buffer between his household and the settlements of early colonists. Part of it was used as a kitchen garden for growing fruits, vegetables and grains for the household and for senior members of the military.

The garden also became a setting for the pleasure, edification and interest of colonists deprived of the amusements they had known at home. The colonists strolled on the lawns to the accompaniment of band music, and, with their familiar landscape designs and well known fruit and flowers, the gardens evoked a nostalgia for 'home'. A piquant interest was often created by the arrival of a new botanical find from an exotic outpost, and many gardens had a zoological section for the display and acclimatisation of wild animals, birds and sometimes even fish, as an added delight.

In the two earliest gardens, those of Sydney and Hobart, there were official visiting hours and restricted entry, permitting the admittance of respectable citizens only, thus the gardens became symbols of law, order and social status. They were places in which to be seen. Only gradually were plots of classified plants included for the study of botany as a science.

In the major gardens there were sections of experimental horticulture where imported plants such as grapes, apples and other fruits, grains, and, in the northern gardens, sugar, spices and fibres could be acclimatised and observed for possible economic use. From these, thousands of seeds and cuttings were eventually propagated and distributed to settlers and farmers.

Many of the ships arriving in an Australian colony carried seed boxes containing seeds packed in either cotton, beeswax or sometimes in sand, and wooden planter cases, containing small plants, which had been strapped onto the poop deck for the voyage. These were nurtured in botanic gardens in Australia.

It was a two-way traffic. On the return journey, the seedboxes and planter cases and, after their in-vention in 1833, the Wardian cases (a type of sealed mini-glasshouse, similar to a terrarium, in which plants could survive for up to eight months without water) carried Australian plant species. The plants were prepared for the voyage in a special section of the botanic gardens.

Even before there were official botanic gardens in Sydney, indigenous plants were being raised for export on the southern slopes of Farm Cove. Europe knew of the *Pultenaea retusa*, the *Hardenbergia violacea*, the *Eucalyptus robusta* and many Hawkesbury sandstone-country plants from New South Wales. These had been sent, probably as seeds, from Sydney to European botanic gardens before the turn of the eighteenth century.

According to Gregor Kraus (1841–1915), professor of botany and director of the botanic garden at Halle, East Germany, there were nine main periods of plant importation into European gardens. During each period, plants from one region were more striking or predominant than those from other regions. The sixth period, from 1771–1820, was the period of imports from New Holland, as Australia was then called, although Australian plants still continued to be imported well after that date.

The interest with which Europeans greeted Australian flora was not surprising. Most of it, particularly from New South Wales, was startling to European eyes and different from any flora they were familiar with in the northern hemisphere. Through Australia's long isolation of approximately sixty million years as a continent, its flora had developed independently from the flora of other parts of the world, even the southern hemisphere. The country was regarded by Europeans as 'a living museum'.

In 1859, Joseph Dalton Hooker, who later became the director of the Royal Botanic Gardens, Kew wrote that the 'flora of Australia has been justly regarded as the most remarkable that is known . . . considered differing fundamentally from those of other lands'. The scientific climate of England and Europe was receptive to it, for the colonisation of Australia occurred at a time of great interest in England and Europe in the natural sciences, particularly botany.

The late seventeenth century had seen the dis-

Specimen of *Actinotus minor* from the Banks and Solander collection, collected during Cook's voyage of 1770. *(Queensland Herbarium, Brisbane)*

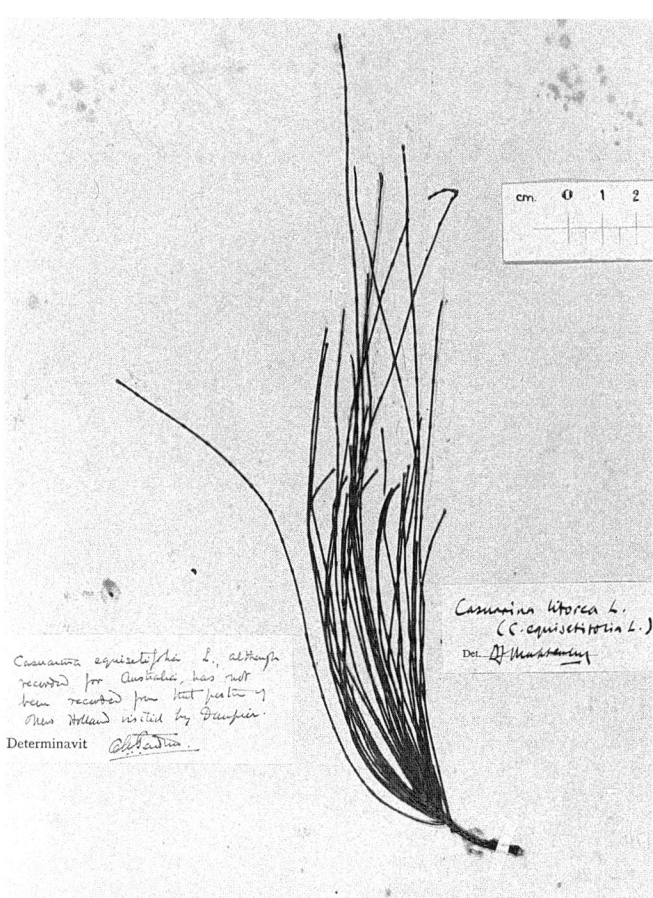

Specimen of *Casuarina equisetifolia* collected by William Dampier during his voyage to New Holland in 1699. *(Fielding-Druce Herbarium, Oxford)*

covery of the botanical riches of the North American lands. By the beginning of the eighteenth century, plants and seeds were being imported by Dr Henry Compton, Bishop of London, Peter Collinson and other plant hunters in Europe.

Then in 1731 British naturalist Mark Catesby published *Natural History of Carolina, Florida and the Bahama Islands*. With its marvellous colour plates of North American trees this publication created enormous excitement in botanical and gardening circles, and by the mid eighteenth century the collecting of exotic plants had gathered momentum. The end of the century brought a wealth of plants from the Cape Province of South Africa with its pelargoniums, ericas, proteas and cape bulbs to add zest to botanical awareness.

Australian flora had first been seen and collected by a Dutch explorer Willem Vlamingh in the Swan River in Western Australia in 1697 and in 1699 by the English buccaneer William Dampier, also in Western Australia. However, it did not become familiar to European botanists and gardeners until the return to England of Captain James Cook in the *Endeavour* on 12 June, 1771.

During the course of the three-year circumnavigation of the world, and the longest sea voyage since that of Francis Drake in 1577, the botanists Joseph Banks and Daniel Carl Solander, a Swedish protegee of the eminent botanist Linnaeus from Uppsala, had amassed three thousand botanical specimens, representing a thousand species.

Shortly after the ship's return, Joseph Banks organised an exhibition of specimens of flora and fauna at his Soho Square house in London for selected men of science and influence. The exhibits, arranged in glass cases designed by Solander, included 412 sheets of herbarium specimens of Australian flora from which 365 finished drawings by artist Sydney Parkinson had been prepared. This was just under half the number of finished drawings of the entire collection, and of the areas represented in the collection Australia provided the majority of drawings. Viewing the dried herbarium specimens side by side with the more realistic drawing of the living specimen gave those attending the exhibition a clear impression of the species on display.

Through the exhibition, and the publication in 1773 of *Cook's First Voyage* by Oxford University Press, in which many of the natural history observations came from Bank's then-unpublished journal, the flora of New Holland became the most desirable of the age and replaced that of the Cape Province in novelty.

The voyage to the South Seas established Joseph Banks as the scientific pundit of the age and in 1771 he was appointed the scientific advisor to George III, and the unofficial director of the Royal Botanic Gardens at Kew.

Specimen of *Clianthus formosus*, Sturt's desert pea, collected by William Dampier at Shark Bay in Western Australia during his voyage to New Holland in 1699. *(Fielding-Druce Herbarium, Oxford)*

The gardens had been started by the King's mother, the Princess Augusta, in 1759 as a small botanical collection and arboretum in 3.6 hectares round her house, 'Kew House', in Richmond, Surrey. They were enlarged by George III in 1771 on the death of his mother when he amalgamated Kew House and its grounds with his own adjoining property, Richmond Gardens.

Joseph Banks's aim was to make the King's collection of plants the most curious, the richest and the largest in the world. He established a policy of sending young botanists and gardeners as 'King's botanists' to gather plant specimens from the new worlds of the Americas, the Cape Province, Africa, China, India and the West Indies, and, after he had recommended it as a penal colony, New South Wales, and subsequently New Zealand and the Pacific Islands.

By 1820, Banks had enlarged the Royal plant collection by many thousands of species, aided by the establishment of the British botanic gardens of Palmplemousses in Mauritius, St Vincent in the West Indies, and those in Jamaica, Calcutta, Penang, Sydney and Trinidad during the latter half of the eighteenth century and the beginning of the nineteenth century.

Joseph Bank's first plant-collector for Australia, George Caley, a young Yorkshire man, arrived in Sydney in 1800. That year, Banks and the new Governor, Philip Gidley King, set up a botanic garden at Rose Hill, near the newly established Government House. The area was later called Parramatta.

Using the botanic garden at Rose Hill as a base for growing and sorting indigenous plants for export to the gardens at Kew and for acclimatising fruit, vegetables, grains and ornamental plants sent by Banks from Kew for the settlers, Caley collected plants and worked from Parramatta for the next ten years.

The Rose Hill botanic garden was also used as a depot by Robert Brown, a young Edinburgh University medical student with a passion for natural history who Joseph Banks had chosen as botanist aboard the *Investigator* during Matthew Flinders's voyage to map the Australian coast from 1801–1803. (Using his inimitable diplomatic flair, Banks had ensured a safe passage from England through French waters during the Napoleonic wars.)

Brown's collection of some 3400 Australian plant species, two thousand of which had not been seen in Europe before, and the illustrations of some of them by his gifted botanical artist, Ferdinand Bauer, accelerated the acceptance of Australian flora in England and Europe. (He is credited with the introduction of many banksias, hakeas, brachysemas and dryandras.)

Robert Brown published the results of his Australian plant collection in 1810 in Part I of *Prodromus Florae Novae Hollandiae et Insulae Van Diemen*. In it he described two thousand Australian plant species. He had collected them in varying stages of development – a new method – and his book helped transform botanical classification and to launch a new study of plant geography.[1] Though it sold only twenty-six copies, it was later described by Sir William Hooker, the director of the Royal Botanic Gardens, Kew as 'the greatest botanical work that has ever appeared'.

Brown's later discovery of what became known as the Brownian Movement in plants, the molecular bombardment of minute particles which oscillate rapidly when suspended in liquid, made him one of the most influential botanists the world has known. As Banks's private secretary and the legatee of Banks's herbarium and library, Robert Brown continued to influence the study of Australian flora and the course of events in early Australian botanic gardens.

Bank's second plant-collector to Australia, Allan Cunningham (1790–1839), arrived in Sydney in 1816, only months after the Botanic Gardens, Sydney had acquired its first superintendent, Charles Fraser. Allan Cunningham had worked at the Gardens at Kew with the superintendent William Aiton since he had been seventeen years of age. He spent fifteen years in Australia, travelling and botanising widely, and introducing thousands of new Australian plant species to Europe. He used the Botanic Gardens,

[1] Anne Moyal, *A Bright and Savage Land.*

Sydney as his base, and often collaborated with Charles Fraser.

Both Banks and King George III died in 1820. By then, with the colonial botanic gardens established, the collection of plants for Kew passed to the resident colonial botanists and to the superintendents of the botanic gardens. The necessity for King's botanists abroad no longer existed.

The demand for flora from the new world in both England and Europe increased after Banks's death. There was a need to fill the gardens of the new rich of the industrial revolution with interesting specimens and to supply the gardens and the glasshouses of the aristocratic plant collectors. The Duke of Devonshire's plant collection at Chatsworth in Derbyshire became famous and his conservatory, with the world's largest glass area, was so large that a horse and carriage could drive down the central aisle.

By 1825 England grew thirteen thousand plant species, a quarter of the estimated plant reserves of the world. Australian flora was particularly sought. The illustrated catalogues of the large London nursery firms, Loddiges and Veitch, with their international trade, promoted the acquisition of flora.

Joseph Banks's successor as the director of the Royal Botanic Gardens, Kew – the first official director – Sir William Hooker, appointed in 1841, continued to take an interest in the nurturing of colonial botanic gardens, including those in Australia. A former protegee of Banks, a surgeon and the professor of Botany at Glasgow University, Sir William was particularly interested in economic botany and encouraged the founding of the colonial botanic gardens of Perendaya, Ceylon, Singapore and Brisbane, which were influential in the establishment of the world tea, rubber and sugar industries respectively.

Though the Australian botanic gardens of Sydney, Hobart, Brisbane, Melbourne and Adelaide were, by the mid 1850s, colonial government institutions, they were influenced both directly and indirectly by the Royal Botanic Gardens, Kew. Sir William Hooker approved or disapproved most senior appointments to colonial botanic gardens, including those of Australia. He collected a remarkable botanical library; substantially increased the number of specimens in the herbarium and the number of living plant species in the gardens since the term of Joseph Banks; and enlarged the general garden from 4.3 hectares (11 acres) to 30.3 hectares (75 acres) of garden, incorporating 109.3 hectares (270 acres) of parkland, to bring the total area of the Royal Botanic Gardens to 139.6 hectares (345 acres). He attracted to Kew the brightest young horticulturists, botanists and botanical artists in Britain; many of them eventually worked in colonial botanic gardens, and of Australia's founding and early curators, superintendents and directors, seven were 'Kewites'.

They maintained strong links with the 'mother garden' in England and many of the original type specimens of Australian flora were still, as they were in Banks's time, sent first to Kew to be assessed and stored, or to the British Museum, as were the original descriptions of the flora. In 1863, Sir William Hooker commissioned George Bentham, the leading British botanist of the day, to compile a *Flora Australiensis*. Hundreds of thousands of indigenous Australian plant specimens were sent to Kew by Ferdinand Mueller, the director of the Melbourne Botanic Gardens, and the first internationally trained botanist to work in an Australian botanic garden. Mueller collaborated with Bentham, enabling him to accomplish the task. Completed in 1878, the *Flora* was a seven volume work of 4200 octavo pages of plant description unrelieved by botanical illustration. It was considered a giant among botanical publications of the day and is still as yet the only complete work comprising all the then-known Australian flora.

Sir William Hooker was succeeded in 1865 by his son Joseph Dalton Hooker, a brilliant botanist who had been assisting his father for ten years. As Kew's new director he continued the Hooker interest in the botanic gardens of Australia, corresponding closely with the directors of the major botanic gardens. Kew was still the centre of a vast botanical network of which Australia was a valuable link. The work of Ferdinand Mueller of the Melbourne Botanic Gardens, Richard Schomburgk of the Adelaide Botanic Garden, Charles Moore from the Sydney Botanic Gardens, and Walter Hill from the Brisbane Botanic Gardens helped make Kew supreme amongst the botanic gardens of the world.

But by the 1870s and the 1880s, the Australian gardens began to emerge as notable entities in their own right, both for their beauty of design and settings and for the scientific expertise shown by their participation in the frequent international exhibitions of the day. Of the Sydney Botanic Gardens, the English writer Anthony Trollope commented: 'Nothing that London has, nothing that Paris has, nothing that New York has, comes near them in loveliness.' International scientific honours and awards were bestowed upon Ferdinand Mueller and Richard Schomburgk particularly, and both were knighted by Queen Victoria. Charles Moore was chosen as a commissioner for the 1867 Paris Exhibition.

Visitors to Australia at this time were often astonished at the number of botanical gardens and parks.

Trollope, who visited the main centres of Australia in 1871, wrote: 'Australians are laudably addicted to public gardens. There are, in all large towns, either in the very centre of them, or adjacent to them, gardens rather than parks which are used and never abused.'

The reason for this he suggested, was: 'The land has been public property and space for recreation has been taken without cost price. In this way a taste for gardens and, indeed to some extent a knowledge of flowers and shrubs has been generated and a humanising influence in that direction has been produced.'

Comparing the gardens to those in Europe, he added: 'Distance is the drawback in the old cities. Parks for the people were not among the requirements for humanity when our cities were built and the grounds necessary for such purpose had become so valuable when the necessity was recognised that it is only with great difficulty that we have been able to create them . . . Even the gardens of the Bois de Boulogne are too remote for daily purposes, but the gardens of Sydney are within easy reach of the combined towns of Sydney and Woolloomooloo.'

No doubt the early nineteenth century political theories on colonisation of British statesman Edward Gibbon Wakefield advocating a new life for the industrial classes had an influence on the formation of public recreation space. He extolled the 'saneness' of living close to the natural seasons and to the land.

By the 1870s and 1880s the landscape 'paradise' style of botanic garden which had stemmed from the eighteenth century landscapes of William Kent, Capability Brown and Humphrey Repton were found throughout Australian public gardens. It was a style, according to English garden historian Edward Hyams, 'superbly expressed in Australian gardens'. In adopting it, Australia's botanic gardens superintendents showed their dependence on the influence of the English thought of the day and particularly on the designs of the gardens at Kew.

That style had begun in George III's reign with Capability Brown's remodelling of Richmond Gardens and had been completed in Sir William Hooker's time with the addition of new parkland, incorporated into a design by W.A. Nesfield which used the new Palm House and Temperate House as features within it.

The sinuous lines of the naturalistic landscape style of Kew accommodated the large numbers of botanical species being sent in from the world's botanical outposts. It was a complete contrast to the previously fashionable seventeenth century French landscapes, which featured rigid formal outdoor architectural spaces made from hedging in which only a limited number of species could be used.

The world popularity for the 'jardin anglais' during the next forty years threatened the already established scientific features of many botanical gardens. 'Landscape gardens are the rage', reported the minutes of a 1910 municipal meeting in Geelong, Victoria. As the controversy over botanical and scientific functions versus those of a recreation park became marked, even the established picturesque style at the Royal Botanic Gardens, Kew, with its integral scientific role, was threatened. Joseph D. Hooker was forced to defend it, and his own position as director, when, in 1873, there was pressure to remake the gardens as a typical London recreation park. A commissioner for works in Gladstone's government, Acton Smee Ayrton, brought the matter to a head and after a debate in parliament in August had resulted in deadlock, only public outcry and the support of Hooker's scientific friends, Charles Darwin and Thomas Henry Huxley, saved Kew as the scientific institution it is today. In Melbourne, as a result of a similar controversy, the scientific, internationally famous formal botanic gardens of Ferdinand Mueller were completely destroyed and reconstructed as a landscape garden by William Guilfoyle.

The landscape style of botanic garden and the function for which it was designed were quite different from the first botanic gardens of Europe which were built in Pisa in 1553–1554, Florence in 1545 and Padua in 1545. For the first time, plants were grown for the collection of scientific data, instead of as medicines in monastic gardens of previous centuries. Arranged with one species to a bed, in brick planter beds built in geometric patterns, the plants were labelled and could be seen, touched, smelt and observed at close quarters. The beds were contained within four distinct squares, later thought to represent the four continents of Europe, Asia, America and Africa, although planting was not always geographical. (The discovery of America, with its wealth of new flora, is thought to be responsible for the establishment of botanic gardens.) Similar gardens were built in Leyden and Montpellier in the sixteenth century.

The seventeenth century brought botanic gardens to Oxford, Paris and Uppsala. These were also built in four distinct squares, but the plants were in straight lines and the gardens were encyclopaedic, able to be read at a glance. The Paris garden, with a mount, had habitat planting.

A century later, the physic botanic garden was the predominant type, and of these the Chelsea Physic Garden, which pioneered the system of plant classification according to sexuality, was after 1722 preeminent in collecting the botanic riches of the world.

But with Joseph Banks, and the botanic brilliance of his superintendent at the Royal Botanic Gardens, Kew, William Aiton, the architecture of William Chambers and the landscaping of Capability Brown, Kew's gardens became the model for the rest of the botanic world.

With the retirement of Joseph Hooker from Kew in 1885, the founding work of the great botanic gardens of Europe and of England and the British colonial world, including Australia, was completed. The prevailing botanic garden ideal of including examples of each of the world's plant species – a sort of Noah's ark concept – was waning as the number of plant finds increased and it plainly became impossible to enclose them in one walled garden.

Within Australia, during the first half of the twentieth century, with two World Wars, the general depression and little funding for botanic gardens, there were no major developments in botanic gardens and existing species were merely maintained. Economic botany and experimental horticulture were taken over from the gardens' staff by the government departments of agriculture and by the CSIRO. The general gardens policy turned to flori-

The lagoon of the Royal Botanic Gardens, Melbourne provided a superb setting for relaxation and enjoyment in the 1860s and 1870s. *(National Library of Australia)*

culture as the world style of public gardening evolved from primarily concentrating on trees and shrubs, to experimenting with bedding plants using new hybrid flowers.

Although herbaria attached to the gardens developed as scientific centres for the study of Australian flora, there was little integration with the living collections within the gardens visited by the general public.

In the last three decades, with the development of interest in the environment, ecology and a general awareness of gardens and plants, there has been a second wave of botanic garden building in Australia. In Perth, the Kings Park and Botanic Garden was opened in 1963 and in Canberra the Australian National Botanic Gardens was officially opened in 1970. Both are major botanic gardens and both specialise in growing Australian flora.

There is today as much interest, locally and internationally, in the wealth of indigenous Australian flora being discovered as there was in the times of the original founding of the botanic gardens. Most of the older gardens, already filled with plants, have opened 'satellite' or annexe gardens usually devoted to the growing of indigenous native flora. The new

Mount Annan Native Botanic Garden, a satellite of the Royal Botanic Gardens, Sydney, is, with its four hundred hectares, one of the world's largest botanic gardens.

New botanic gardens specialising in indigenous flora have recently begun in Port Augusta, South Australia, Mount Isa, Northern Queensland, and at Orange in western New South Wales.

Staff of the major botanic gardens are contributing to a new *Flora of Australia* to replace George Bentham's *Flora Australiensis*. Eight of the estimated fifty volumes have been published and it is expected that the project will take another twenty years to complete.

Herbaria in all states have become a vital part of botanic gardens and in Sydney, Melbourne, Adelaide and Canberra, they are contained within the gardens and are integrated with the living plant collections. The other state herbaria are administered by state government departments and located separately from the gardens. In all state botanic gardens, indigenous plants threatened with extinction are being propagated and 'saved' and if possible, their habitats studied with a view to their preservation.

The palm house of the Adelaide Botanic Gardens during the 1870s. *(Adelaide Botanic Gardens Collection)*

The rose garden of the Adelaide Botanic Gardens, designed by Richard Schomburgk, in its infancy in 1867. *(Adelaide Botanic Gardens Collection)*

For all the emphasis on the scientific and educational role of Australia's botanic gardens, their function as pleasure gardens has not been overlooked and is being enhanced through increased funding. A world-wide interest in garden styles which has developed over the past forty years has led, for example, to the making of Japanese gardens within the botanic gardens of Rockhampton and the Royal Tasmanian Botanical Gardens. New garden architecture, in the form of revolutionary glasshouses funded by the Bicentennial Authority, will add interest to the botanic gardens of Adelaide, Wollongong and Sydney.

An international awareness of the importance of the garden as part of a nation's cultural heritage has emphasised the value of the designs and flora of Australia's nineteenth century botanic gardens. In Victoria, as part of the state's 150th birthday celebrations in 1985, funds were made available for the rejuvenation of twenty-six regional botanic gardens. And in all botanic gardens in Australia, the trees and shrubs of the nineteenth century – the *Ficus* species of the fig tree avenues which are found in the gardens from Townsville in the north, to Adelaide in the south; the jagged araucarias: the hoop pines, bunya bunya pines and the Norfolk Island pines; the immensely tall palms; the flowering tropical shade trees of the northern botanic gardens; and the dignified exotic conifers of the southern ones – are being preserved and treasured as symbols of the Victorian age. The existing ferneries, conservatories, bandstands, pavilions, statuary and gates are also being preserved and restored.

In the early botanic gardens, lawns are still being maintained. In the new ones, they are being planted to ensure that the chatting, the strolling, sitting, picnicing and reading and the enjoyment of the seasonal flower displays within them continues. For, as Joseph Maiden, the director of the Royal Botanic Gardens, Sydney, 1896–1926, concluded: 'This is how the majority of visitors occupy their time. They take their botany mildly.'

1

THE ROYAL BOTANIC GARDENS, SYDNEY

The Royal Botanic Gardens, Sydney, including the National Herbarium of New South Wales, cover 30 hectares. They are set amidst a 33.5 hectare parkland comprising The Domain, the Art Gallery of New South Wales and the grounds of Government House, and the whole area lies between the elegance of Macquarie Street and the harbour front of Farm Cove and Woolloomooloo Bay. It is acclaimed world-wide as a superb site.

The Gardens have other aesthetic advantages. There are huge slabs of sandstone breaking from grassy slopes, and grotesque rock shapes forming small escarpments – remnants of the original terrain; an undulating landform that gives harbour views often incorporating the Sydney Opera House; a gardenesque planting that offers romantic walks and thick pockets of subtropical shrubbery; and there is the curving beauty of a sandstone sea-wall lapped by the harbour.

The principal botanical feature of interest is the Palm Grove, established in the 1860s and added to since then. It was considered by Sir Edward Salisbury, director of the Royal Botanic Gardens, Kew, who visited Sydney in 1949, to be one of the finest in the world. Edward Hyams, in Sydney in 1968 to do research for his book *The Great Botanical Gardens of the World*, regarded the palm collection as the world's second finest after that in the Fairchild Tropical Garden of Florida. Today other Australian botanic gardens have more than Sydney's estimated 110 species, but none has a more beautiful nor as old a palm grove. The Gardens also have a fine collection of trees from the South Pacific and the other Pacific regions and many *Ficus* species of great size.

The climate supports a wide range of flora and there are approximately eight thousand labelled species throughout the Gardens. Cold-climate species like rhododendrons do not flourish, however. In January the maximum temperature averages 25.7 degrees C and the minimum 18 degrees; in July there is an average maximum of 17.1 degrees and a minimum of 8.3 degrees. The average annual rainfall is 1216 mm, with no frosts.

Two gaps in Sydney's collection are being filled by botanic gardens elsewhere in New South Wales. The Mount Tomah Botanic Garden in the Blue Mountains, which opened in November 1987, will specialise in cold-climate flora and has been built as a joint Bicentennial project of the New South Wales and federal governments. The limited variety of Australian native plants in the Royal Botanic Gardens will be supplemented by flora to be planted in a new annexe of the Gardens on the outskirts of Sydney near Campbelltown, the Mount Annan Botanic Garden, which is a Bicentennial project of the New South Wales Government.

The indigenous flora of the Royal Botanic Gardens, Sydney has mostly disappeared from the site. As the ground is swampy, plants would have included *Banksia integrifolia*, rushes, the swamp oak *Casuarina glauca*, and *Eucalyptus tereticornis*, two specimens of which remain on the lawn in the Lower Garden near Mrs Macquarie's Road. There are some fairly old *C. glauca* near the bronze lion sculptures and these are believed to be suckers from the original vegetation.

The superficial impression given by the Gardens is one of a mannered late nineteenth-century pleasure park, and from photographs taken in the 1870s and 1880s it is easy to see why the Botanic Gardens of Sydney were considered, even on a world scale, to be important at that time. With their abundant statuary, elegant pathways, interesting trees (including many tall araucarias) and fine gates, they prompted Anthony Trollope, visiting in 1871, to write in his *Australia and New Zealand*:

For loveliness and that beauty which can be appreciated by the ignorant as well as by the learned the Sydney Gardens are unrivalled by any that I have seen. The nature of the land, with its green slopes down to its own little bright bay has done much for them and art and taste combined has made them perfect.

In those days visitors entered the Gardens from the city, through Bent Street, close to the original Government House, across Macquarie Street and down Fig Tree Avenue, an august line of Moreton Bay fig

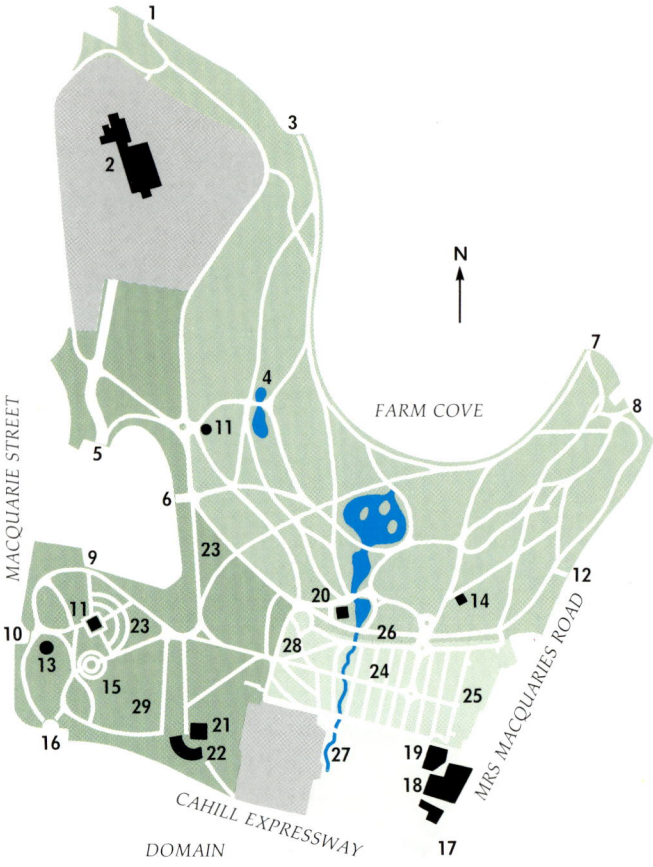

KEY TO MAP

1 Opera House Gate
2 Government House
3 Queen Elizabeth II Gate
4 Twin Ponds
5 Government House Gate
6 Rose Garden Gate
7 Yurong Gate
8 Victoria Lodge Gate
9 Conservatorium Gate
10 Palace Garden Gate
11 Shelter Pavilion
12 Henry Lawson Gate
13 Phillip Fountain
14 Maiden Memorial Pavilion
15 Pioneer Memorial Garden
16 Morshead Foutain Gate
17 Woolloomooloo Gate
18 National Herbarium
19 Visitor Centre, Education
    Section and Maiden Theatre

20 Gardens Restaurant and
    Kiosk
21 Pyramid Glasshouse
22 Arc Glasshouse
23 Rose Garden
24 Plants, Evolution and Man
25 Succulent Garden
26 The Macquarie Wall
27 Botanic Gardens Creek
28 Palm Grove
29 Palace Garden Grounds

▢ Lower Garden

▢ Middle Garden

▢ Upper Garden

trees, *Ficus macrophylla*. The gates, new in 1873, set on ornate sandstone pillars, were of wrought iron and painted mauve, with the Royal Cipher and the Crown outlined in gold. The entrance gave a sense of occasion to arrival at the Gardens. Once at the gates, visitors turned left and went down a pathway of 100 yards (91.4 m) to the Main Walk in the Upper Garden, a path that led towards the Lower Garden and the harbour. Before entering the Lower Garden, visitors would stop at the 'Wishing Tree', a Norfolk Island pine, *Araucaria heterophylla*, almost 100 feet (30 m) high; they would circle it three times in each direction and make a wish. Supposedly one of the oldest trees in the Gardens, the Wishing Tree is believed to have been planted by Mrs Macquarie in 1817. It was removed in 1945.

Today, with the severing of the Royal Botanic Gardens from The Domain by the Cahill Expressway, built in 1962 for traffic to and from the eastern suburbs and the North Shore, the original entrance no longer exists. What remains of the fig trees in the Fig Tree Avenue are stranded in the middle of the Expressway, and the main gates have been moved along the street and around the corner to create a new entrance, the Woolloomooloo Gate, in Mrs Macquarie's Road. But although the atmosphere of a grand formal entrance has been partly lost, the new entrance is still the best way to start a tour of the Gardens and to appreciate their history, development and plants.

You commence by walking through the park of grassy slopes, specimen trees and shrub beds that is today's Upper Garden. Here is a section of plants from South Africa and, alongside the administration buildings, herbarium and Visitor Centre, a garden of Australian native plants that were discovered by Dr Daniel Solander, the Danish botanist who accompanied James Cook in the *Endeavour* and landed at Botany Bay in 1770. Where the ground becomes flatter, you enter the former Upper Garden and today's Middle Garden and descend the Main Walk between twin *Livistona chinensis* palms planted by Governor Bourke in 1847, to the sandstone Macquarie Wall, built between 1813 and 1816. On either side of the Main Walk are narrow beds separated by paths. The beds run from north to south towards the harbour, and on the right-hand side are filled with large old trees and underplanted with shrubs and groundcovers. On the left-hand side the beds are planted with an educational display of herbs and medicinal plants. The main axes in the Middle Garden run from east to west in three main avenues. As you reach the end of the Main Walk, you pass through the Macquarie Wall into the Lower Garden, which begins with a thick, irregular planting of semitropical and tropical trees and spreads out in a gardenesque manner over an informal apron of land towards the sea.

Bisecting the Gardens, from the Upper Garden through the Middle Garden to the Lower Garden, is Botanic Gardens Creek, which descends through six weirs to become pools for statuary, ducks and water plants in the Lower Garden. The water – stormwater runoff from The Domain – is recirculated, but it is not constant enough to be useful for irrigation. It is, though, significant in a decorative sense. The creek also has historic importance as it was, with the Tank Stream (just under a kilometre away to the east), the source of the first fresh water found by Captain Arthur Phillip when he arrived at Port Jackson as governor of New South Wales with the First Fleet of convict settlers on 26 January 1788.

Within a month of landing, Governor Phillip asked his personal servant Henry Edward Dodd to establish a small farm in the thin pockets of alluvial soil on the banks of the creek. (A separate garden of fruit and ornamental seeds was begun at Government House in Bridge Street, less than half a mile

Above: **The Palace Garden in 1888, near the Conservatorium of Music.** *(Tyrell Collection, Museum of Applied Arts and Sciences, Sydney)*

Below: Fig Tree Avenue, originally the main entrance to the Botanic Gardens until 1962 when the Cahill Expressway was built. *(National Library of Australia)*

Left: The main gates of the Gardens were made in Sydney in 1873 and moved to Mrs Macquarie's Road in 1962.

Right: Ficus macrophylla, the Moreton Bay fig, planted in the Lower Garden about 130 years ago, is the broadest tree in the Gardens.

to the west of the creek.) Convicts cleared the land round the creek of rushes, *Banksia integrifolia* and sword grass and by June of 1788, 9 acres (3.6 ha) of corn, wheat and other cereals had germinated and were growing.

By September of that year there were 20 acres (8.1 ha) of barley, wheat and other grains growing on the east side of the creek – cultivated in furrows running east to west, the direction of today's three main pathways in the Middle Garden. But the crops did not flourish. The soil of Sydney is deficient in phosphorus and so poor in most nutrients that today a tomato cannot be brought to fruiting stage in the Botanic Gardens without fertilisers. Governor Phillip thought there was probably more fertile land to the west of Sydney, and a river too. In May of 1788 he took a party up Sydney Harbour and, following a branch of the harbour inland, discovered the site for a town he later named Parramatta. There was 'a very fine run of fresh water, the country on both sides pleasant and the ground apparently fit for opening with far less trouble than in any other part of the Harbour and with the soil good'. By November, after a second visit, Phillip had marked out a town, a farm, and the site for a Government House to serve as a country retreat from Sydney. He called the area Rose Hill after George Rose, the British secretary to the Treasury, and by 1789 the farming venture had moved from Farm Cove to Rose Hill.

Up until this time all the botanical acclimatisation of plants sent from England for the settlers, and of indigenous plants thought to be either useful or ornamental, for export to such places as the Royal Botanic Gardens, Kew, had taken place at Farm Cove. In 1792 Governor Phillip marked out a 'Governor's Demesne' from the head of Darling Harbour to the head of Woolloomooloo Bay, 'to be reserved for the Crown and for the Use of the Town of Sydney'. Inside this he marked a second line from Government House to the head of Woolloomooloo Bay in which the Government Farm and the subsequent Botanic Gardens area was enclosed. This became the Inner Domain of 54 acres (21.9 ha), while the Outer Domain was 125 acres (50.6 ha). But the fertility of the soil was so low at Farm Cove that the botanical acclimatisation work as well as the farming was removed to Rose Hill, and in 1800 the new governor, Philip Gidley King, announced to Sir Joseph Banks the forming of a 'Botanic Garden' at Rose Hill just over the river from Government House and George Caley, the official Collector of Plants for Sir Joseph Banks, went to live there. Governor King wrote to Banks: 'I have fixed Caley at Parramatta to which place he gives the preference and I hope he will begin in earnest. I have marked out a Botanic Garden . . . It is already receiving plants and Caley has the use of Government House to dry his specimens.'

For the next ten years the Botanic Garden at Rose Hill was the official botanic garden for Sydney, where imports were acclimatised, indigenous plants established and seeds sorted before the voyage back to England. A duplicate of all seeds and plants sent by ship to England and Europe was grown and kept at Rose Hill to prevent irrevocable loss in the event of shipwreck on the risky voyage home.

When George Caley returned to England after ten years at Rose Hill, the garden there probably deteriorated. But during Governor Lachlan Macquarie's term of office the gardens round Government House, Parramatta, developed considerably and Allan Cunningham, the King's Collector for Kew Gardens, wrote to Sir Joseph Banks in 1818 that 'Macquarie is forming a Botanic Gardens at Parramatta'. This was a subsidiary to the Farm Cove gar-

den, for by 1816 the major botanic garden for Sydney had reverted to Farm Cove on the eastern side of Botanic Gardens Creek where the first farm had been.

Governor Macquarie had a wall built by convicts round the edge of the Inner Domain to protect the incipient botanic garden from sea winds and trespassers. Outside it, a road was built to culminate in a loop at Ansons Point where there was a large slab of rock facing the harbour. This became known as Mrs Macquarie's Chair and the road, which delineated a favourite walk of the governor's wife, became Mrs Macquarie's Road. The road measured 3 miles and 377 yards (5.17 km), and when it was finished on 13 June 1816 the garden within the wall – today's Middle Garden – was dedicated as the Sydney Botanic Gardens.

Later that year Charles Fraser, a 28-year-old member of the 46th Regiment, arrived on the convict ship *Guildford* and was appointed to the position of superintendent by Governor Macquarie. Born in Scotland, Charles Fraser may have had horticultural experience in the Royal Botanic Garden of Edinburgh. It is also thought that he worked in the gardens of the Duke of Norfolk, but his exact experience

of gardening before he came to Australia is not known.

One of his first tasks as superintendent was to gather seeds of the 'choicest' native plants of New South Wales for the Emperor of Austria, whose collection of plants at Schönbrünn in Vienna was the only one in Europe to rival that of George III at Kew. In 1817 Charles Fraser was unofficially appointed colonial botanist on the expedition of explorer John Oxley to trace the course of the Lachlan River. In the hills around the new town of Bathurst Fraser went botanising with Allan Cunningham, and he must have benefited from the association.

In 1819 Charles Fraser shipped three cases of 'rare and choice plants' from further expeditions with John Oxley to Lord Bathurst, the colonial secretary in England, and throughout his term as superintendent, Charles Fraser travelled extensively to collect plants, visiting New Zealand, Norfolk Island, Perth, Tasmania, many outlying parts of New South Wales and, with Allan Cunningham, to Brisbane, where he founded the garden that became the Brisbane Botanic Gardens.

Charles Fraser's duties, outlined in 1821 when he was also officially recognised as colonial botanist, were to cultivate exotic plants introduced into the Gardens and distribute their offshoots to colonists; to collect and send indigenous plants to institutions and individuals who had contributed to the collection of plants in the Gardens; to propagate interesting crops, fruits and grasses; and to make a general collection and 'arrangement' of plants.

There was still dissatisfaction over the state of the soil at Farm Cove, and in 1821 Charles Fraser persuaded Governor Macquarie to try a new site 2 miles (3 km) away at Double Bay, where 15 acres (6.1 ha) were cleared and fenced on the western slopes of the bay ready for planting a 'Colonial Botanic Garden'. In 1825 the plans were abandoned by Governor Brisbane, Lachlan Macquarie's successor, who added 5 acres (2 ha) to the Farm Cove garden instead of continuing the Double Bay project.

Charles Fraser persevered in his attempts to establish the Gardens at Farm Cove and by 1831 they were organised into four sections: the Kitchen Garden, for growing vegetables for the governor and other officials, in what is now the Upper Garden; the Fruit Garden, on the eastern side of Botanic Gardens Creek, where the first farm had been and which subsequently became the Middle Garden of today; the Botanic Experiment Garden on the western side of Botanic Gardens Creek – also in today's Middle Garden – where the first grape vines in Australia were planted, exotic new species were

Below: Nymphaea capensis, a South African water lily.

Opposite: The Main Pond in the Lower Garden with a *Yucca elephantipes* from Central America in the foreground.

tried, and indigenous flora was acclimatised for export; and the Lower Garden, which extended to just below the present Gardens Restaurant and had become a place for strolling, with curving paths and tropical and subtropical trees. Elizabeth Macarthur, wife of the pioneer John Macarthur, wrote of the Lower Garden: 'It will be very beautiful. The introductions from Moreton Bay promise to be very ornamental.' (Today the seeds planted by Charles Fraser from his mission to Brisbane in 1828 have grown into some of the loveliest trees in the Gardens.) In all, the Gardens had grown to 24 acres (9.7 ha).

By 1831 Fraser had fulfilled all his duties as laid down by Governor Macquarie. The Gardens supported three thousand species of food plants, crops and ornamental plants and thousands of specimens had been propagated from them and given away to colonists and institutions, with the public being urged to collect their cuttings at the Gardens. Fraser was so popular that one Sunday in 1829 the gates of the Gardens had to be locked because of the vast number of people overwhelming him. He had experimented with the grape vines imported by James Busby and by 1829 had 543 varieties flourishing in the Gardens. He had made from poppies some 'excellent opium' and distilled oil from eucalypts for the medical profession. Crops, including cotton seed, had been supplied to dependencies such as Norfolk Island. Fraser's collections of indigenous plants and seeds had been sent throughout the world to institutions and to individuals who had provided new species for the Sydney Gardens. The rock orchid, *Dendrobium speciosum*, gigantic lily (or Gymea lily), *Doryanthes excelsa*, and the Norfolk Island pine, *Araucaria heterophylla*, were particularly sought after and were exported to England and Europe. In 1830 he supplied seeds and cases of living plants to Glasgow, Edinburgh, Liverpool, Batavia and Mauritius and received plants from China, Ceylon, Calcutta, Madagascar, Mauritius, the Cape of Good Hope, Marseilles, Paris, the Royal Botanic Garden of Edinburgh, Liverpool Botanic Gardens and many other institutions. He was important to botany in his day and was mentioned 220 times by George Bentham in his *Flora Australiensis*.

Fraser's salary had begun at £91 5s 0d a year and progressed to £150 a year, out of which he had to pay for all plants and seeds to be sent abroad. In fact, when he died in December 1831 at the age of forty-three, he was in debt.

After Fraser's death, the colonists were concerned that the new superintendent should be a man of science and the colonial secretary Alexander Macleay offered the position to Allan Cunningham, the King's Collector, who had arrived back in England in 1831 after seventeen years away, fifteen of them spent in Australia. Cunningham declined, but proposed his brother Richard, who had worked as a librarian in the office of Kew Gardens for twenty years. Richard was also recommended by William Aiton and Robert Brown. He accepted the offer

Right: Yuccas from Central America and a jacaranda in the Lower Garden.

Below: Brachychiton rupestre, the Queensland bottle tree, is one of the distinctive trees of the Gardens.

when it was made to him and was in turn accepted by the colonists.

Arriving in Sydney in 1833, on the same ship as some of James Busby's vines, Richard Cunningham set about tidying the Gardens. His brother Allan had already written a memorandum to Governor Bourke suggesting a policy that included the introduction of a classified system of plant arrangement; a herbarium to be related to the living plants; an annual report; and a regularly printed catalogue of the plant collection.

In 1833-34 the new superintendent distributed thousands of plants grown in the Gardens, including 3075 grape vine cuttings. He sent seeds and plants to England and Europe and travelled on collecting missions to New Zealand, Norfolk Island and Tasmania. In March 1835 he joined as botanist the expedition of Major Thomas Mitchell down the Bogan River in New South Wales, and a month later wandered off from the camp one day and was not seen again. It is believed that he was killed by Aboriginals as he wandered, dazed and lost, into their camp.

Cunningham's brother Allan accepted the post of superintendent and colonial botanist when it was offered once again, but he too had insufficient time to make any tangible impact on Sydney's Botanic Gardens. From the moment he arrived in February 1837 he was restricted by the Committee of Superintendence of the Australian Museum and Botanical Garden, a committee of influential citizens including Alexander Macleay and William Macarthur and set up by Governor Bourke. The Botanic Gardens subcommittee stipulated Cunningham's duties and took control of most of the Gardens land. The superintendent's authority was to be confined to the Kitchen Garden and the Inner Domain.

After being ordered to train convicts as gar-

deners, as well as to concentrate most of his activities on the Kitchen Garden, Allan Cunningham resigned in December of 1837. 'Tell all that I have discharged the Government cabbage patch in disgust', he wrote to friends in England, 'and am now to enter . . . on a more legitimate occupation for a few months'. He went to New Zealand as a freelance botanist but, already ill with tuberculosis, returned to Sydney and died within a few days in a cottage at the Botanic Gardens, on 27 June 1839. He was forty-seven years old.

A period of maladministration, drought and difficulty followed for the Gardens until the Committee of Superintendence appointed John Carne Bidwill as director in August 1847. Bidwill, an English botanist and naturalist who had been in Australia and New Zealand since 1838, had collected, among other indigenous plants, the bunya pine, *Araucaria bidwillii*.

Bidwill's appointment was endorsed by Charles FitzRoy, the governor, but simultaneously the colonial secretary in London, Lord Grey, authorised the appointment of Charles Moore, sponsored by Dr John Lindley, chairman of the Royal Horticultural Society. The London appointment took precedence over that of the governor of the colony and John Bidwill was dismissed from office on 1 February 1848.

Charles Moore, aged twenty-seven and arriving in Sydney in 1848, was appointed to the position of director of the Sydney Botanic Gardens at a salary of £300 a year which had been fixed on the understanding that John Bidwill would be given the position, and was reduced to £200 after Moore became director. Moore's brief from the committee was that the Gardens should be concerned with the latest methods of horticulture and botanical science as well as being a pleasant place of resort for Sydney's citizens. The committee had already stopped the production of fruit and vegetables for officials, and had prohibited the distribution of free crops, fruit trees and the seeds of ornamentals from the Gardens.

Trained in botany and horticulture in the Botanic Gardens of Trinity College in Dublin, by the curator, Dr James Townsend Mackay, Charles Moore had worked at Regent's Park in London and the Royal Botanic Gardens at Kew. His brother David, also a trained botanist and horticulturist, became director of Glasnevin, the National Botanic Garden of the Irish Republic, Dublin.

According to Dr Lionel Gilbert, author of *The Royal Botanic Gardens, Sydney: a history 1816–1985*, Charles Moore's first year in Sydney was productive. He renovated paths throughout the Gardens and built new ones in the Lower Garden. Using stone from the demolition of Old Government House in Bridge Street – a new Government House had been built adjoining the Gardens at Farm Cove between 1837 and c.1845 – he organised construction of a sea-wall along the harbour frontage of the Gardens land. Called the Thirty Years Wall, it took that length of time to complete. Labour, since the emancipation of convicts, was always a problem, as was the paying for it, and the gold rushes of New South Wales and Victoria, early in the 1850s, claimed most casual labour.

Charles Moore started a classified botanical ground in the Lower Garden. In the Middle Garden there were beds showing plants for medicinal, arts, manufacture and domestic uses. He found the design of Charles Fraser and the Cunninghams satisfactory and reported to parliament in 1848 that 'The general good taste displayed in the original design

Above: Botanic Gardens Creek in the Middle Garden.

Below: Botanic Gardens Creek with the Gardens kiosk in the background.

of the gardens has rendered it unnecessary to effect any alteration in this respect'.

He went plant-collecting to the South Seas, and to the Blue Mountains and the Myall Lakes of New South Wales. He began lectures on botany, for medical students and the general public, that continued until the 1880s. Plants and seeds, thanks to his connections in England and Europe and his closeness to his brother at the Glasnevin Gardens, came to the Sydney Botanic Gardens from all over the world, and indigenous plants were exchanged for them. Construction began on the ponds in the Lower Garden, and a water supply for the upper part of the Gardens was helped by the building of a water pipe

from Macquarie Street. One of the nine men working at the Gardens at this time (the late 1850s) for 6s 6d to 7s 6d a day was, according to Dr Gilbert's research, Anthelme Thozet, a trained botanist from Paris who later went on to Rockhampton, Queensland, and helped to establish the Botanic Gardens there.

By 1860 the reclamation from the sea in the Lower Garden was well under way. Swampy land had been filled with street sweepings, silt, and offal from slaughterhouses and was then grassed. A long walk was made around the inside of the sea-wall. Trees planted then – the Moreton Bay fig, *Ficus macrophylla*, the Mexican bald cypress, *Taxodium*

mucronatum, the palms, *Phoenix reclinata*, which together with the araucarias distinguish the skyline around the Gardens – are the giants of today. The Palm Grove, near Middle Garden, was started in 1862. The tallest tree in the grove, the Queensland kauri, *Agathis robusta*, was planted in 1853 and was already large in the 1860s.

Charles Moore also thinned out trees already planted in the Lower Garden. Of his planting policy he wrote:

It is not by the acquisition of a vast number of species or the crowding together of an endless variety of plants possessing neither beauty or value that public taste is improved or the cause of education served but rather by a judicious selection and cultivation of such types of genera as will, by properly illustrating natural families at the same time, be interesting and instructive. Plants therefore remarkable for beauty, singularity or their utility for man are the most suitable for this purpose and such it has been my objective to select for and retain in this establishment.

Summerhouses, including an Oriental-style pavilion with a thatched roof of *Xanthorrhoea*, appeared in the Gardens, and in 1860, after much lobbying from

the public, an aviary in green and white was opened where today the Succulent and Cactus Garden is situated on Mrs Macquarie's Road. This was decorated with statuary consisting of a pair of Egyptian sphinxes and a pair of lions. Two years later a zoo was incorporated and remained until it was removed to Moore Park in 1883. The aviary was retained until 1940.

By the 1860s Charles Moore was well in control of his Gardens, responsible to the governor through parliament. The Committee of Superintendence begun by Governor Bourke had been relieved of its duties by Governor FitzRoy in 1851. In 1855 a recommendation that Moore be demoted from Director to Curator and made subject to the control of three commissioners was quashed by the governor, Sir William Denison.

Up until this time Charles Moore had been dispensing plants raised at the Gardens to nurseries, private citizens and botanic gardens overseas – particularly popular indigenous plants like the waratah, the silky oak, the cypress-pines (*Callitris*), the hoop pine and the Christmas bush. He was prohibited from supplying plants that local nurserymen could distribute, and came into dispute with the prominent growers Thomas Shepherd and Michael Guilfoyle, the father of William Guilfoyle, who designed the Melbourne botanic gardens.

Moore's catalogue of plants in the Gardens was published in 1857. It listed three thousand species of flowering plants and ferns, including 750 indigenous to New South Wales and 110 from other Australian states. A new report in 1871 showed that the Upper Garden had the greatest number of species.

Collecting trips into the hinterland of New South Wales continued, for there was an international appetite for news of the economic possibilities of new species, particularly timber trees. For the London Exhibition of 1862 Moore collected 115 samples of 'wood indigenous to the Northern Districts of the Colony', and for the Paris Exhibition of 1867 he again gathered samples of New South Wales timbers.

In 1878 it was suggested that an International Exhibition be held in Sydney, the first in Australia. To be sponsored by the Agricultural Society of New South Wales, it would feature works of art and industry. The New South Wales colonial architect James Barnet drew plans for a Garden Palace to house the exhibition on land between the Government House stables – the present Conservatorium of Music – and Governor Bourke's statue, an area then used for grazing animals.

The Garden Palace had a dome 100 feet (30.5 m) in diameter under which was a statue of Queen Victoria; four towers; and a floor area of over 8½ acres (3.4 ha). It was the epitome of the High Victorian style and the surrounding gardens complemented it. Landscaping by the Botanic Gardens staff involved the making of an 'instant' garden of twenty-eight thousand seedlings of bedding plants, lawns and shrubberies in nine months. The building was officially opened by the governor, Lord Loftus, on 14 September 1878. Three years later, on 22 September 1882, the Garden Palace burnt down during the night. The exhibits, including the library of the Linnaean Society, was destroyed with the building, as were the gardens nearby. When the debris was removed, there were 19 acres (7.7 ha) of cleared land available to be given to the Botanic Gardens. Today

this section, along the Macquarie Street boundary, is 'The Palace Garden'.

Charles Moore had already added to the Gardens area during the 1870s by demolishing the convict barracks built in Allan Cunningham's time, and the old glasshouses in what was originally the governor's kitchen garden to make the grassed landscape of today's Upper Garden.

As well as being responsible for the Botanic Gardens and the relatively untouched Domain, Moore was in demand as a garden-maker. He worked on Hyde Park, and the grounds of Government House, Admiralty House, the University of Sydney and the Art Gallery of New South Wales and, in 1888, the new Centennial Park.

Moore was also active as a botanist. In 1883 he wrote a *Census of Plants of New South Wales* as a ready reference to all the plants found either naturalised or indigenous in New South Wales. In 1893, with the help of Ernst Betche, a botanical collector on the Gardens staff, he published the *Handbook of the Flora of New South Wales*, which remained the definitive work on the flora of the state for the next sixty-five years and was the only work describing its vascular plants. Within the Gardens, plants were labelled to show their natural order, scientific name

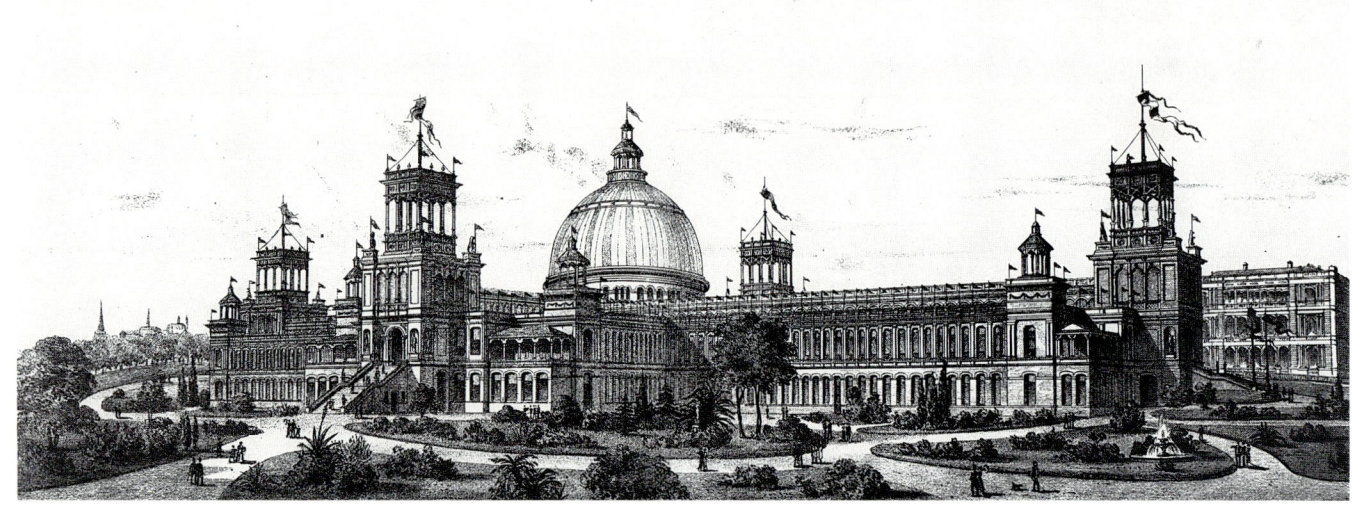

and authority, common name and the native country of each plant. The same system is in use today.

When Charles Moore was first appointed to the Gardens, there was not one dried specimen in the herbarium. Early specimens collected by Charles Fraser and the King's Collectors had all been sent back, often in duplicate or triplicate, to the Royal collection at Kew or to the British Museum (Natural History). The start of a Sydney-based herbarium came with the gift to Moore in 1852 of four packets of specimens of tropical and subtropical flora collected by the surveyor-general, Major Thomas Mitchell. These specimens had been arranged and described in England by Dr John Lindley, Sir William Hooker, George Bentham and Professor De Vriese. Charles Moore had added to the collection. Yet when Moore retired from his position as director of the Botanic Gardens and as colonial botanist in 1896 at the age of seventy-six, only a dozen portfolios of 150 pages each were found.

Moore had enlarged the 'living' collection of plants in the Gardens twenty times since his arrival in 1848 and the area of the Gardens by many acres. He had been president of the Royal Society of New South Wales, a Fellow of the Linnaean Society of London and of the Royal Horticultural Society, and a member of the Royal Botanical Society.

The new director, Joseph Henry Maiden, was an Englishman like Charles Moore. A science graduate from the University of London, he arrived in Sydney in 1880 intending to make only a visit. He took a position as curator of the Museums collection of the Technological, Industrial and Sanitary Museum of New South Wales, which eventually became the Museum of Applied Sciences and organised the Museum's collection of specimens exhibited in the Sydney International Exhibition of 1879, a display that took up an acre (.4 ha) of space in the Garden Palace. After the fire of 1882 Maiden gathered up what was salvageable from the collection and exhibited it in a nearby pavilion in The Domain. It became the official museum, incorporating an herbarium.

Through Joseph Maiden's enthusiasm a new museum was built at Ultimo, and he became the superintendent of the Museum of Applied Sciences in 1894. By this time he had educated himself in botany by attending Charles Moore's botany lectures and by forming a correspondence with Ferdinand Mueller in Melbourne and a friendship with the pioneer botanist and teacher the Reverend William Woolls, of Parramatta. He joined the Linnaean and Royal Societies and studied botany in the bushland around Sydney, becoming particularly interested in the genus *Eucalyptus*. On the retirement of Charles Moore, Joseph Maiden was appointed Director of the Botanic Gardens with an annual salary of £515 and a house worth £105 a year. Dr Lionel Gilbert writes of Joseph Maiden: 'Possessing a width of training, interest and vision not shared by any of his predecessors, he had the additional advantage of nearly fifteen years' experience in the public service, which had sharpened his wits without reducing his energy or enthusiasm'. He prepared annual reports and attended to detailed matters such as adequate lighting, seating, lavatories, drinking fountains and pathways. He lit the 'Sea Walk' in the Lower Garden, and incorporated 5 acres (2 ha) from the Outer Domain into the Lower Garden, making the ring of waterfront of the Lower Garden complete. In 1901 he made sure that the remains of Allan Cunningham were transferred from the Devonshire Street cemetery to a memorial obelisk that stands in the Gardens near today's Garden Restaurant. He also drained and sewered the Gardens.

Left: **The gates of the Palace Garden in Macquarie Street depict the outline of the Garden Palace.**

Bottom left: **A drawing of the Garden Palace built in 1878 and burnt down in 1882.** *(Tyrell Collection, Museum of Applied Arts and Sciences, Sydney)*

Right: **The Macquarie Wall, which separates the Middle Garden from the Lower Garden, was built by convicts from 1813 to 1816.**

Left: The Royal Botanic Gardens, Sydney looking towards Woolloomooloo Bay from a Macquarie Street building.

Above: Dendrobium speciosum, the rock lily, on a sandstone escarpment in the Lower Garden near Farm Cove.

Above right: View of the Gardens from the Opera House side of the Gardens looking towards Mrs Macquarie's Road.

One of his main preoccupations was to establish a reputable herbarium, to stock it properly and to get increased funds from parliament to do so. He recruited Ernst Betche, the German-born botanical collector who had assisted Charles Moore, to help him as 'Chief Botanical Assistant'.

In December 1899 a newly built herbarium and library, museum and administrative centre was opened (the Visitor Centre in the Gardens today). At the opening ceremony Joseph Maiden announced that 'Scientific Botany now has its headquarters in New South Wales. A botanic garden cannot properly perform its functions without the support of a rich herbarium – a garden of dried plants'. He set out to collect indigenous species throughout New South Wales. In a few years, travelling to Lord Howe Island, to the Snowy Mountains and throughout the state, he had added several thousand specimens to the herbarium collection. (Later, in 1905, during a visit to England, he sought some specimens of the Banks-Solander collection from the Botany Department of the British Museum and was given 586, which are among the most prized specimens in today's collection of over a million.)

Maiden's contribution to the Botanic Gardens was primarily his role of shepherding the Gardens into the public consciousness. He frequently spoke at public meetings, particularly those of the Royal Society. His botanical writings included *A Critical Revision of the Genus Eucalyptus* and *Forest Flora of New South Wales*, and were illustrated by Margaret Flockton of the Botanic Gardens staff. When he retired in 1924, he had established the Gardens on a sound scientific basis.

There was little change in the Gardens during Dr G. P. Darnell-Smith's term of directorship, which followed that of Joseph Maiden, nor during the earlier stages of that of Robert Anderson, the first Australian-born director of the Sydney Botanic Gardens and an agricultural science graduate, who was appointed in 1936. Standards declined at one stage during this time, however, due in part to reduced funding throughout World War II.

Physical disruption of the Gardens area occurred in 1956 when the City Council of Sydney took land in the Outer Domain below the Art Gallery of New South Wales for the construction of a car park. Forty-seven comparatively rare tree specimens were lost. In 1958, plans were announced to build the Cahill Expressway through the Gardens and The Domain. Fig Tree Avenue was destroyed and twenty-four palms and a dozen other trees were lost. But the greatest loss was that of the unity of the Botanic Gardens with The Domain, which had adjoined each other since Governor Phillip's reservation of the parkland in 1792. Director Robert Anderson, disillusioned by the planning of the Expressway, wrote: 'If there had been in the first place full information as to the extent of the destruction contemplated by the planning authorities, it would have been possible to offer other suggestions for the route of the Expressway'.

Scars left by the excavations for the Expressway were covered by new plantings. The new entrance in Mrs Macquarie's Road was landscaped with pools and prostrate plants. A new entrance was made in Macquarie Street using the gates that had been commissioned after the fire at the Garden Palace, depicting in their design the dome of the Palace. This entrance, landscaped into Macquarie Street, leads to the Palace Gardens.

Robert Anderson's successor, Knowles Mair, a botanist at the Sydney Gardens who had formerly been curator of the Darwin Botanic Gardens, inherited the task, in 1969, of completing the reconstruction after the building of the Expressway and the Domain Parking Station. Australian plants were extensively used and a new section of New Zealand native plants was begun near the State Library of New South Wales. Today this is an attractive section of the Palace Gardens with many varieties of *Hebe* and a fine kaka beak, *Clianthus puniceus*. One of Knowles Mair's special interests was the glasshouses, which were decrepit and not keeping pace with modern technology. Many were removed, and Mair instigated the designing and building of the new Pyramid Glasshouse, erected in 1970–71. The first of its kind in the world, it contains an internal staircase so that visitors can observe all levels of tropical growth as they wind through it.

In 1968, during Knowles Mair's period as director, the herbarium became part of the Botanic Gardens. Knowles Mair retired in 1970, and in 1978, at the suggestion of the subsequent director, Dr Lawrie Johnson, the administration of the Royal Botanic Gardens and the National Herbarium of New South Wales was handed over, after seventy years with the Department of Agriculture, to the Premier's Department of the New South Wales Government. (The epithet 'Royal' had been added to the name of the Gardens in 1959, on the recommendation of the trustees.) In 1980 a Bill for a Royal Botanic Gardens and Domain Trust Act was passed in parliament. This sought to prevent any further erosion of the Botanic Gardens and Domain land, which since 1916 had diminished from 72.6 hectares to 63.04 hectares. Also in 1980, Centennial Park, which had been administered by the staff of the Botanic Gardens since Charles Moore's directorship ninety-two years before, became autonomous.

Dr Lawrie Johnson, a botanist who was an authority on the genus *Eucalyptus*, was the director of the Botanic Gardens from 1972 to 1985. He had been on the staff of the Gardens for twenty-five years before becoming director. He oversaw the building of a new herbarium, which was opened in 1982. Costing $4 million, it was named in honour of Robert Brown, the botanist who had collected so many Australian native plant specimens during his voyage with Matthew Flinders from 1801 to 1803. This linked the buildings of the Botanic Gardens: the former herbarium, which was opened as a visitor centre for exhibitions and book sales and named the Anderson Building, and the former director's residence, the Cunningham Building, constructed during Charles Moore's directorship in the 1870s and now used by administrative staff.

Dr Johnson proposed the 'thematic' planting scheme that is evident today in the Gardens. The rainforest trees, many of which had been collected from the wild by the early directors Charles Fraser, Richard and Allan Cunningham and Charles Moore and are scattered throughout the Gardens, have been supplemented by a new section of tropical and subtropical rainforest flora near the Pyramid Glasshouse.

The palm collection, which is planted in three different groves in the Gardens, has been thinned of duplicate species, particularly in the original Charles Moore Palm Grove, and new species have been planted. The *Ficus* collection, which is concentrated mainly in the Lower Garden, has been rationalised and is centred on the slope below Government House, with many additional species planted. The plantings of native Australian species indigenous to the Sydney region have been strengthened by a garden bed of local species adjoining the Cunningham Building in the Upper Garden, and the long bed of native small trees and shrubs along the boundary between the Gardens and Mrs Macquarie's Road has been thickened with new plantings. A collection of eucalypts on the lawns of the Lower Garden outside the Macquarie Wall, has been underplanted with a new collection of cycads, many of which came from the original palm grove and are already mature.

Another interesting addition for visitors wanting to see Australian plants is the Myrtales Bed near the Twin Ponds in the Lower Garden which contains members of the order Myrtales. This includes members of the family Myrtaceae, to which the largest number of Australian native plant species, including *Eucalyptus*, belong. The planting will show the relationship between the Australian members of the order and those from other parts of the world.

The new Succulent Garden, built on the site of the original aviary of Charles Moore, features a large collection of succulents and cacti that have been gathered together from throughout the Gardens and were part of earlier collections. Roses will have a new garden, soon to be built, showing the place of the rose in history, species of particular historical significance, and species of old-fashioned roses. The present collection features only modern roses. The educational role of the Gardens has been extended by the recently established section in the Middle Garden called 'Plants, Evolution and Man', with its selection of herbs, medicinal and dye plants. Ferns will be accommodated in a new glasshouse, currently being built, which will be linked with the existing Pyramid Glasshouse. Within these two glasshouses it is hoped to grow half the world's fern species.

A comprehensive new *Flora of New South Wales* is being published, and a review of the classification of the genus *Eucalyptus* by Dr Lawrie Johnson, Ken Hill and Don Blaxell is in progress.

But perhaps the most exciting development for the Royal Botanic Gardens, Sydney is the opening of the satellite gardens Mount Tomah Botanic Garden in 1987 and Mount Annan Botanic Garden in 1988, under the aegis of the current director, Professor Carrick Chambers.

The Mount Tomah Botanic Garden, 120 kilometres west of Sydney, a 31 hectare high-altitude garden on one of the basalt caps of the Blue Mountains, was donated to the Royal Botanic Gardens in 1972 by Effie Jane Brunet who, with her husband Alfred Louis Brunet, a nurseryman and horticulturist, raised flowers for market there for thirty years.

The Garden, which developed slowly from 1972 until 1982, when $2.5 million granted by the com-

Top right: **The Pyramid Glasshouse in the Palace Garden.**

Right: **The Palm Grove flanked with cliveas.**

monwealth and state governments turned it into a Bicentennial project, opened on 1 November 1987. As Mount Tomah includes the Brunets' original garden, it has an established atmosphere. There is also a tranquil framework in the remaining stands of tall brown barrel trees, *Eucalyptus fastigata*, which is a species endemic to the basalt caps of the Blue Mountains. Inside the property you find remnants of forty-year-old hedges of laurel, *Prunus laurocerasus*, which defined the paths and roadways of the earlier garden. There are also tall and mature rhododendrons, which surround a newly built complex of buildings arranged round a service courtyard.

A formal garden, framed by the laurel hedges,

around the car-parking area at the top of the plateau, has been established and includes a rosarium featuring old-fashioned roses, a lawn terrace with a herbaceous border, a pergola supporting cold-climate climbers, a medieval herb garden, and a paved area with sundial. The canopy specimen trees include the deciduous conifer the dawn redwood, *Metasequoia glyptostroboides*, a sycamore and some *Quercus* species. Round the depot buildings, a bank of *Rhododendron ponticum* forms the backdrop for a thick, formal planting of cold-climate shrubs. Arboreta will sweep outwards from the depot in three main sections, consisting of plants from North America, including birches, beeches, conifers and

Below: The fruit of *Encephalartos altensteinii*, a cycad species from South Africa.

Right: *Archontophoenix cunninghamiana*, the bangalow palm, and some cabbage tree palms frame the obelisk memorial to Allan Cunningham, the King's botanist who died in a house in the Gardens in 1839.

Bottom: One of the palm groves in the Gardens features a central *Phoenix dactylifera*, the date palm.

oaks; Eurasia, including plants from India, China and Japan, with many magnolia and rhododendron species; and the southern hemisphere, embracing Australia, New Zealand, South America, South Africa and montane Africa.

Between the Southern Hemisphere and Eurasian arboreta, in a natural gully just below the new Visitor Centre, a large basalt rock garden with a waterfall, cascades, large ponds and a bog garden, with many varieties of water plants and sedges of the southern hemisphere is being built. A scree slope in the complex, with pockets of different soils to get as wide a range of cold-climate alpines and rockery plants of the southern hemisphere as their habitats and the climate will allow, has also been built.

The excellence of the soil (volcanic, over sandstone) and climate (rainfall averages 1535 mm) can be gauged from the garden surrounding the house of the resident superintendent, horticulturist Tony Curry, which began in 1978 and is already a full-blooming English-style country garden with lawns, feature trees and large flowering shrubs.

As Mount Tomah Botanic Garden lies in an area subject to bushfire, a fire-retardant garden along the north-west boundary by the road has been planned; it will include trees and low-growing plants. The formal terrace garden was designed principally by Geoffrey Britton of the New South Wales Government Public Works Department.

The Mount Annan Botanic Garden, a 400 hectare former dairy farm just off the Southern Freeway between Camden and Campbelltown and about an hour and a half's drive from Sydney, is the newest addition to Sydney's Royal Botanic Gardens, and was designed by the Gardens staff and staff of the New South Wales Public Works Department.

The land, chosen late in 1984 from five areas of crown land proposed for a satellite garden to specialise in Australian flora, as a state government Bicentennial project, was a unanimous choice by the committee of staff from the Royal Botanic Gardens. Mount Annan has good topography, with an undulating form that can permit many microclimates. Its position is good for the quick growth of plants, having a north–south orientation and many ridges to the east and west. However, it mainly faces east, and is protected from the westerly winds, which in this part of outer Sydney are harsh and either very cold or very hot. The low rainfall, of 700 mm, is approximately 25 per cent lower than that of Sydney and, combined with a somewhat wider temperature range than Sydney's, produces a climate in which native plants from all over Australia can grow. The elevation of the land, which reaches 197 metres on the top of Mount Annan, also helps to create a wide variety of climatic conditions.

The view from Mount Annan is far-reaching, with a foreground of farming country which has been settled since John Macarthur built Camden Park in the 1830s. Further on are the housing estates of Camden and Campbelltown. As well as giving the Royal Botanic Gardens a native plants annexe, the Mount Annan garden will provide an inviolate parkland for one of Sydney's fastest-growing regions.

Pockets of indigenous vegetation remain. This is mainly forest red gum, *Eucalyptus tereticornis*, grey box, *E. moluccana*, and narrow-leaved ironbark, *E. crebra*. There is also a remnant of dry rainforest on the upper part of Mount Annan. Eventually the area will be linked by 9 kilometres of one-way loop road, with many provisions for stopping and picnicking. Twelve kilometres of walking tracks have been planned. Irrigation for the property will be provided from a 300 mm ring main and the bottom of the U-shaped valley forming the core of the Garden will have two manmade lakes, which will help to irrigate the land.

A large terrace garden covering 4 hectares is planned. It will contain a sequential display of each order and family of the Australian flora, from the most primitive to the most advanced. Around the lakes, plants from particular Australian families will be grown in appropriate terrain and climatic conditions. There will also be a conservation area featuring rare and endangered native plants and a lakeside garden of water-loving plants.

The arboretum featuring Australian trees, including as many of the 850 species of *Eucalyptus* as the climate will support, is already being planted. It is hoped that the new Garden will become an important centre for research into the horticultural needs of Australian plants.

With the addition of its satellite gardens, the Royal Botanic Gardens of Sydney will become nationally important as a fine botanical collection. Already they have a unique blend of history, situation and beautiful old trees. The best time to see the Gardens is in the spring – late September – when the collection of flowering trees from all over the world is blooming. Beginning in the Upper Garden, you come upon the *Dombeya rotundifolia*, a small old tree smothered in pale pink fragrant blossom, which is the dominant feature of the bed of South African native plants. This tree and other large old trees in the Upper Garden were planted as part of the original garden surroundings of the original superintendent's house occupied by Charles Fraser and Charles Moore. In the Middle Garden, where you can still see a couple of old citrus trees remaining from the Fraser–Cunningham times, when this section of the Gardens was used to acclimatise and experiment with fruit trees, a selection of small flowering trees from all over the world includes *Rothmannia globosa*, the tree gardenia from South Africa, with its white fragrant blossom. Specimens of *Magnolia* x *soulangiana*, planted on either side of the Main Walk, are full of flowers in the spring, and to the left of the Main Walk, near the Palm Grove, you can see the tree waratah, *Oreocallis wickhamii*, brought by Charles

Overleaf: The Levy Fountain in the Upper Garden commemorates Lewis Wolfe Levy, MLC in 1888.

Top left: The current Rose Garden in the Palace Garden.

Bottom left: Delphiniums and a tamarisk tree in the small beds of the Middle Garden.

Right: Pansies in the Middle Garden, looking towards the skyscrapers of Macquarie Street.

Fraser and Allan Cunningham from Brisbane; it has bright red waratah-like blooms on the ends of its branches. An *Erythrina acanthocarpa*, from Africa, is a brilliant coral tree, also in the Middle Garden. An African thorn tree, *Acacia karoo*, with yellow flowers, can be seen among the rather haphazard planting of the same area. Coming to the Macquarie Wall, you find the Azalea Walk, running east to west towards the Palm Grove, closely planted with large white, mauve and pink Indica azaleas, many of which were already notable in Charles Moore's day, the 1860s and 1870s. Against the only remaining glasshouse in this section of the gardens is a bank of *Rhododendron ponticum*, probably planted in the 1860s and the only rhododendrons to do well in the conditions of the Gardens.

Near the ponds in the Lower Garden there is, in early spring, the startling white blossom of *Pyrus pashia* from Afghanistan to western China, which, unlike other members of the pear family, holds its blossom for at least three weeks. Generally, though, the Lower Garden is notable for its evergreen trees of significant shapes and sizes. One particular well-loved tree is the lilly-pilly, *Syzygium francisii*, near the Gardens Restaurant. Another is the fine example of *Araucaria cunninghamii*, probably planted by Richard Cunningham and named after his brother Allan. Many of the old trees in this section of the Gardens are unlabelled because their origins and exact name are undocumented and still unknown today. Whether the visitor is interested in such botanical matters or is content to stroll through the Gardens for relaxation and pleasure, it is impossible to escape their atmosphere of history and age. Wrote Joseph Maiden: '. . . in Sydney we thickly coat the botanical pill with the sugar of "garden of pleasure"'.

2

A REGIONAL GARDEN IN NEW SOUTH WALES

WOLLONGONG BOTANIC GARDEN

There are a few places in Australia where rain-forest meets the sea. The Daintree River area in northern Queensland, where the forest plummets 1350 metres down to the Coral Sea, is one. Wollongong in the Illawarra District, an hour's drive south from Sydney, where the southernmost pockets of warm-temperate rainforest grow down to the Tasman Sea, is another.

Some of the Illawarra rainforest can be seen in the Wollongong Botanic Garden, a modern garden opened in 1970 and developed to display plants from all over the world in their natural habitats. Tucked into a small valley, the indigenous rainforest contains most of the famous south coast timber trees, including the red cedar, *Toona australis*, and the coachwood, *Ceratopetalum apetalum*. There are also the decorative blueberry ash, *Elaeocarpus reticulatus*, and the bonewood, *Emmenosperma alphitonioides*, with their brilliant summer fruits; figs including the Illawarra fig, *Ficus obliqua*, the Port Jackson fig, *F. rubiginosa*, the sandpaper fig, *F. coronata*, and the Moreton Bay fig, *F. macrophylla*; the bangalow palm, *Archontophoenix cunninghamiana*, and the cabbage tree palm, *Livistona australis*; *Melaleuca styphelioides*; and many ferns. A small unnamed creek runs through the bush, cascading and meandering over sandstone rocks, and forming pools and torrents in a fashion that excites the envy and admiration of landscape architects.

This is the only part of the garden in which the habitat and the planting are entirely natural. The other plantings – dryland, wet sclerophyll and dry sclerophyll, woodland, the bog garden, the moraine, and the new exotic rainforest section – have been introduced to suitable existing locations, or else the landform has been changed by earth-moving equipment to accommodate it satisfactorily.

Habitat planting is not new to most Australian botanic gardens, particularly the more recently developed ones. You can see it at the Australian National Botanic Gardens in Canberra, with their rock garden and misted rainforest sections, while

there are some forms of it at Mount Coot-tha Botanic Gardens in Brisbane. In Kings Park and Botanic Garden, Perth, plants from some sections have been grouped according to the habitat they favour, just as they have been in the palm grove at the Royal Botanic Gardens of Sydney, the temperate rainforest section of the Royal Tasmanian Botanical Gardens, and the fern gully at the Melbourne Gardens. But no Australian botanic garden has developed habitat planting to the stage reached at Wollongong Botanic Garden.

Wollongong's climate, which is basically monsoonal, will support a wide range of flora from tropical to temperate zones – an advantage in habitat planting. In summer the temperature rarely goes above 35 degrees C and in winter rarely below 7 degrees, and there is an annual rainfall of 1250 mm, falling mainly in the summer. Winters are frostless and not cold enough for the brilliant colouring of autumn foliage, but there is a total of approximately ten thousand species throughout the Garden, with equal numbers of Australian and exotic plants.

The topography of the Wollongong Botanic Garden is ideal for habitat planting. Comprising about 27 hectares of undulating landscape on the western outskirts of the city, the Garden provides the opportunity for creating many microclimates. The highest hill has been mounded to form a desert landscape for dryland plants from inland Australia and succulent plants from dry areas of the world like South Africa and Central America. With gravel-filled gullies, and studded with enormous sandstone 'floater' boulders from a demolition site in the city, some of which weigh 20 tonnes, the drainage in the area is excellent. This helps to compensate for the high rainfall of Wollongong, which is fundamentally difficult for the Mediterranean-climate dryland planting. Although quite recently planted, the hillside is already silver with the foliage of saltbush from South Australia, varied by flashes of mauve from the three varieties of native hibiscus planted there, and the red, yellow and orange flowers of the emu-bushes (*Eremophila*) from the drylands of South Australia and Western Australia. Miniature acacias flourish, and there is mulga from inland Australia flowering – a rarity on the coast of New South Wales.

Left: Mount Keira stands sentinel over the Wollongong Botanic Garden.

On a flat, elevated boundary of the Garden opposite the University of Wollongong grows a border of *Eucalyptus* forest of the wet sclerophyll type, with thickly planted trees, an understorey of tree ferns, and a dense groundcover of *Cissus antarctica*. This type of eucalyptus forest is endemic in the Illawarra region, although here the planting has been introduced. Alongside it a dry sclerophyll forest of widely spaced eucalypts has been planted with an understorey of grevilleas, hakeas, banksias, grasses and herbs.

The lower valley of the site contains an area of woodland planted mainly with exotic northern hemisphere trees including maples, magnolias, dogwoods and birches, among stands of indigenous turpentine trees, *Syncarpia glomulifera*, of immense size. The groundcovers are those seen in the northern hemisphere – primroses, violets, Canterbury bells and foxgloves – and paths are soft with leaf mould and dropped bark. Nearby is an azalea bank with many varieties of kurume and indica azaleas, and camellias scattered through them include one of the best collections of species camellias in Australia, with about thirty species flourishing.

Also in the low-lying area close to the indigenous rainforest and the creek are the bog garden and the moraine habitat, which emulates the suspended lake system of the Illawarra Escarpment and

the perched lakes or tarns formed by glaciers in alpine areas. The director of the Garden, Deane Miller, made many trips to the Snowy Mountains for reconnaissance and seed-collecting, with the result that a seeping, gravelly landscape, where celmisias, helichrysums, trigger plants, mosses, sedges and reeds thrive, has been constructed.

Appointed Controller of Parks and Gardens for Wollongong in 1977, Deane Miller gained a horticulture diploma at the Royal Botanic Garden, Edinburgh, one of the oldest and most prestigious botanic institutions in Europe, and he was profoundly influenced by the curator there at the time, Edward Kemp, a Scot who advocated habitat planting for botanic gardens. Miller was also indirectly influenced by Edinburgh's rock garden, which was made in 1875. With a scree slope, moraine section and thousands of alpine and rock-loving plants, it became the model for important rock gardens throughout the world.

Explaining his planting policy, Miller says: 'Plants do not thrive naturally in isolation. They grow in natural little communities to provide shelter for each other. If in a botanic garden you can create a genuine habitat for plants the visitor can appreciate the total experience of them.' Miller was landscape architect for the Mount Coot-tha Botanic Garden in Brisbane, and during his six-year term there studied landscape architecture at the Queensland Institute of Technology, a course that was an asset in his development of the relatively new Wollongong Botanic Garden. His aim with the land has been to study the site seriously, sometimes choosing an apt habitat after five years of thought.

Not all the planting falls neatly into habitat groupings. On the gentle hillside sloping down to the main valley of the garden, which is exposed to fierce westerly winds that can be either hot or cold, there is a large section of landscaped trees, shrubs and lawns. Set in curving lawns, the plants are graded down from windbreak trees to groundcovers, and are arranged in generic or family groups. Explained Deane Miller, 'With the limited resources of the Garden you cannot have great collections so it is better to show the breadth of the family in its varying forms'. The flowering trees and shrubs section is divided into fourteen families including Lamiaceae, with its *Teucrium* forming the tall breakwind and sloping down through lavenders, rosemaries, Jerusalem sage and salvias to ground level. Other shrubs planted for shelter on the slope, including oleanders and cotoneasters, tie in with the plant families they represent.

There is also a section of landscaped English garden within a wall, with circular beds of modern roses, a pergola, and an herbaceous English border

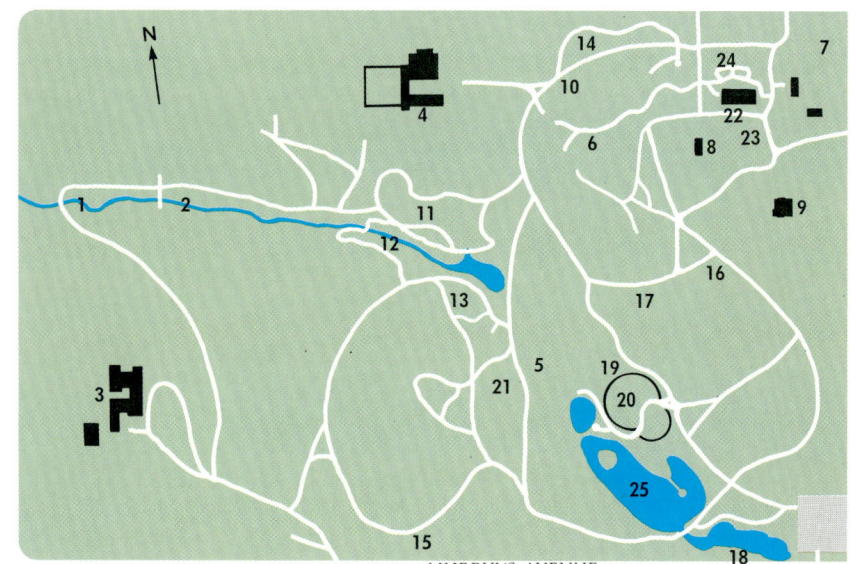

NORTHFIELDS AVENUE

N

ROBSONS ROAD

MURPHYS AVENUE

KEY TO MAP

1 Exotic Rainforest	14 Wet Sclerophyll Forest
2 Subtropical Rainforest	15 Conifers
3 Gleniffer Brae	16 Flowering Trees
4 Nursery	and Shrubs
5 Azalea Bank	17 Herbs
6 Dryland Plants	18 Marsh Plants
7 Plant Houses	19 Perennials
8 Fountain	20 Rose Garden
9 Education and	21 Woodland Garden
Visitor's Centre	22 Dry Tropics
10 Dry Sclerophyll Forest	23 Wet Tropics
11 Illawarra Bog Garden	24 Temperate Regions
12 Illawarra Rainforest	25 Lake
13 Moraine	

with clipped lawns and garden seats. This adjoins the ornamental lake, which contains waterlilies, ducks, and an Oriental pavilion created in 1966 by damming the creek that runs across the garden. From here you look up to the flat-topped escarpment of Mount Keira, which dominates the Garden in much the same way that Mount Wellington overlooks the Royal Tasmanian Botanical Gardens in Hobart. Round Mount Keira and, in the distance Mount Kembla and Brokersnose, the clouds and sky form a kaleidoscope of changing weather, and in this Garden it is the sky, rather than the ground, which first draws the eye.

There are manmade vistas too. At the entrance opposite the university in Northfields Avenue you look up through a tunnel of thick planting to the water display of the Mercury Fountain. This, with the *Jubaea* palm at the Geelong Botanic Gardens, is one of the most dramatic vistas to be seen in Australia's botanic gardens. The fountain, designed by Sydney's Robert Woodward, an internationally known fountain sculptor, was commissioned by Illawarra Newspaper Holdings Pty Ltd to mark 125 years of newspaper publishing in Wollongong. It was meant to stand in the centre of the city, but the building of a mall necessitated the fountain's removal, and Deane Miller found an apt spot for it in the Garden. Here, from the top of the hill, you can look to the coast and catch a glimpse of the sea. Looking in the other direction from the dryland mound, you can see on the horizon the Tudoresque mansion Gleniffer Brae, built in 1928 and landscaped into a garden of rockwork and conifers designed by a Norwegian-born Sydney landscape architect, the late Paul Sorenson.

The house and garden were built for Sidney Hoskins, a founding director of Australian Iron and Steel, which had been transferred, as the Hoskins Steelworks, from Lithgow to Port Kembla near Wollongong in 1926, owing to the good port facilities and the presence of excellent coal in the locality. In 1928 Sidney Hoskins bought several farms beneath the Mount Keira escarpment to build his home,

Gleniffer Brae. In 1954 he donated 25 acres (10.5 ha) of the land furthest from the house to the Wollongong City Council to use as a war memorial park or botanic garden.

The Council's choice was for a botanic garden, but planning and the estimated £160 000 needed for its development were not immediately forthcoming. In 1963 a landscape consultant for the Garden, Professor Peter Spooner of the University of New South Wales, was asked to submit a plan and report on the project. He proposed that the plants should be grouped according to their countries of origin. The areas were Australasia, Indonesia, Malaysia, the Pacific islands, Europe, India, Africa, China and Korea, and the Americas.

Early in 1963 the Engineer's Department of the City Council of Wollongong appointed William Mearns as the curator of Parks and Gardens, to be in charge of the new Garden. The first planting, fencing, clearing, drainage and water reticulation took place in 1964. In 1968 the Wollongong Botanic Garden was opened to the public on Sundays, and by 1969 $85 000 had been spent on its development. There were four thousand plants in the Garden and two thousand in the nursery. Six gardeners, who were studying horticulture at the Wollongong Technical College, were employed. The New South Wales Government's grant of $7500 for the Captain Cook Bicentennial in 1970 provided half the funds needed for a glasshouse, and on 26 September 1970 Mr Sidney Hoskins officially opened the Wollongong Botanic Garden. On 23 November they were opened to the public on a regular basis.

But the growth of the plants was disappointingly slow, although thousands of azaleas flourished in the woodland valley. The soil was heavy and clay-based, a factor that limited the variety of species possible, while the winds, unhindered by today's shelter belts of trees, decimated the new planting on the slopes. Even now the only fine trees in the Garden are the indigenous turpentines and the trees in the rainforest valley.

In 1976 Gleniffer Brae, which had been bought

Top: A walled garden contains a rose garden and an English-style herbaceous border.

Middle: The Mercury Fountain near the entrance of the Garden was designed by internationally known Sydney architect and fountain-designer Robert Woodward.

Bottom: A Chinese-style rotunda in the lake in the Garden.

by the Sydney Church of England Girls Grammar School as a boarding establishment, sold 6.4 hectares of the surrounding land to the Wollongong City Council to add to the botanic garden area, and in 1978 Gleniffer Brae itself and all its remaining land were sold to the Council, more than doubling the original garden area and bringing the total to 27 hectares. The new land for the Garden included the pocket of Illawarra rainforest and, as Deane Miller had been appointed to manage the Garden the previous year, the decision to change the planting policy was not hard to make. Gleniffer Brae became the Wollongong Conservatorium of Music, and the Sorenson Garden, the nucleus for the Garden's conifer section. Deane Miller began the habitat planting by reinforcing the eucalypt groves at the Northfields Avenue entrance with more trees underplanted with ferns and vines. The extant Illawarra rainforest was thickened with indigenous planting and the dryland mound, bog garden, moraine and woodland systems were made.

The form of the new gardens overlaid that of the original, with some concrete paths eliminated and a new pathway system installed. A walking trail with a 2 metre wide path was developed to link the principal collections of the gardens so they could be seen without the visitor having to double back on his tracks. Loop paths 1 metre and 1.5 metres wide were also developed to lead people through specific plant habitats. Plants were labelled, and the flowering trees and shrubs section established.

Recent developments in the Garden include the preparation of a habitat for exotic rainforests covering 3.75 hectares. A windbreak of black wattle, *Acacia mearnsii*, and sweet pittosporum, *Pittosporum undulatum*, has been planted round the perimeter, and a cover crop of bleeding heart, tobacco plant and brown kurrajong planted to provide a closed canopy for the yet-to-be-planted rainforest trees beneath them. This rainforest will take between ten and twenty years to develop. As the rose garden within the walled English garden is not a success,

Top left: Some of the Illawarra rainforest, a natural piece of vegetation within the Botanic Garden.

Above: Illawarra rainforest habitat features timber trees, palms and ferns.

Left: The Joseph Banks Glasshouse, built to commemorate the Captain Cook Bicentenary, holds desert plants, cacti and succulents.

Right: Park-like planting at the entrance to the Botanic Garden.

due to the high humidity within the wall, a new rose garden growing old roses is being planned around Gleniffer Brae. The gardens round the house are to be conserved in accordance with the original planting and design of Paul Sorenson.

Additional habitats available to the Wollongong Botanic Garden are found in the two outstations of the Mount Keira Nature Reserve and the Puckey's Nature Reserve. The former contains a fine patch of Hawkesbury sandstone flora, with all the amazing floral wealth of the Hawkesbury environment. Accessible by road, the reserve will have many native plant species labelled, and the bush enriched with indigenous growth. Puckey's Nature Reserve, a tract of land of 32.5 hectares on the coast north of Newcastle, offers a coastal habitat with its range of salt-tolerant planting, riverine rainforest and heathland planting. Existing flora will be studied and preserved for posterity.

New buildings for the Botanic Garden have been designed by avant-garde Sydney architect Glen Murcutt, and will be built near the existing Sir Joseph Banks Planthouses near the Northfields Avenue entrance. Here are some of the finest turpentine trees in the Garden, and they have been incorporated into the design. A grant of $100 000 for this development, which will include a visitor centre, a herbarium and a new glasshouse, has been made by the Bicentennial Authority. The new tropical display glasshouse will contain a collection of tropical rainforest plants to complement the collection of aquatics and smaller rainforest plants in the Sir Joseph Banks Planthouses. Built in 1970 this glasshouse was reorganised in 1980 to house Australian desert plants and cacti, and is linked with a slatted shadehouse for temperate plants that require protection in Wollongong's climate. With the new addition, the existing glasshouses will form a comprehensive conservatory section for the Garden.

The newest of Australia's pleasure park botanic gardens, Wollongong offers one of the most beautiful and dramatic of their landscapes, limitless space for expansion, and a chance for visitors to see gardens from areas far from Wollongong, by means of the habitat planting.

3

ROYAL TASMANIAN BOTANICAL GARDENS, HOBART

The Royal Tasmanian Botanical Gardens are, at 42 degrees latitude, the most southerly in Australia. As they cover only 13.5 hectares, they are the smallest of the states' botanic gardens. They are the second-oldest in the country, having been started, it is believed, in 1818, two years after those of Sydney. And they seem, even today, to be the most English of the major nineteenth-century Australian botanic gardens.

Their situation at Pavilion Point on the sheltered southern shore of the River Derwent and just over 3 kilometres from the heart of Hobart was considered by nineteenth-century visitors to be 'beautiful'. Looking over gentle slopes running down to the river and then across it to bold eucalypt-covered hills on the opposite side, visitors were reminded of the Scottish lochs.

In 1852 Mrs Charles Meredith wrote about these Gardens in her book *My Home in Tasmania*: 'They are – and I repeat my oft-used term of praise – English looking gardens; not rich in glowing oranges and scarlet pomegranates and golden loquats, not with the gorgeous blossoms of India, but full of sweet homely faces and perfumes.' There was then a great collection of roses displaying over seventy varieties; Tasmania, as it does today, grew the best roses in the country.

The famous Arthur Wall, named for and built under the supervision of Governor George Arthur in 1829, marks the boundary between garden and thoroughfare. Constructed of the red bricks characteristic of Tasmania on the Gardens side, and sandstone blocks on the street side, it is hollow, and was originally heated, like walls in English kitchen gardens, with charcoal-burning fireplaces strategically built in along it. The wall supported fruit trees and climbing plants of 'varied form and radiant bloom'. It was the only 'hot wall' of any size in Australia, and although unheated today still supports semi-tropical creepers like bougainvillea, tecomaria and akebia. A second great wall, the Eardley-Wilmot Wall built by Governor Eardley-Wilmot between 1843 and 1846, added to the atmosphere of English-style enclosure. Of mellow brick, 280 metres long and 4 metres high, it was supported on one side by flying buttresses, and today is believed to be the longest convict-built wall in Tasmania.

When Mrs Meredith wrote of the lawns in the mid-nineteenth century they were 'verdant' from copious watering from a reservoir built halfway down a hillside in 1848, and they linked groups of trees and shrubs planted in the fashionable gardenesque style of the nineteenth century. Planted lavishly throughout were the dark pines and firs beloved during the Victorian era, providing the Gardens then and now with the best collection of conifers to be seen in Australia. They made an interesting skyline against Mount Wellington, the hunchback mountain that dominates Hobart.

By 1850 Hobart Town was a place that, according to Peter Bolger's book *Hobart Town*, could be loved. The Botanical Gardens, visited that year by some 9100 citizens who rang a bell and signed a visitor's book before entering, were part of the city's established charms.

Visitors approached the Gardens through the Queen's Domain, which was laid out with pleasant drives through groves of eucalyptus trees and black wattles, and they passed Government House, completed in 1857 from sandstone quarried in the grounds. With seventy rooms and fifty chimneys, it was acknowledged, according to the English writer Anthony Trollope, who visited Tasmania in 1872, as being the best government house of any British colony. The building rested then, as it does today, in green paddocks studded with groups of oak trees, and it overtopped the walls and hedges separating the governor's grounds from the Botanical Gardens – offering a romantic presence and, at times, a pleasant vista from the Gardens. Today visitors approach the Gardens by the same route. Once past Government House the road, which is planted with giant conifers on both sides, seems to merge with the planting in the Gardens seen behind the Arthur Wall.

At the entrance, the gates of which were imported from England in 1874, it is the conifers, of

Left: The lily pond in the Royal Tasmanian Botanical Gardens, originally built in 1848 as a reservoir.

all the plants, that predominate. There is a giant Spanish fir, *Abies pinsapo*, on one side of the entrance, and on the other a pair of giant sequoia, *Sequoiadendron giganteum*, known to live three thousand years. Also in the upper section of the Gardens is a dawn redwood, *Metasequoia glyptostroboides*. Discovered in China in 1947, it has deciduous foliage that turns red in autumn. Also in this section is a fine example of the Douglas fir, *Pseudotsuga menziesii*, from western North America, which produces oregon timber, and some huge examples of Californian redwood, *Sequoia sempervirens*, the tallest tree in the world. A variety of mature cedars, cypress and pines is also dominant in this area and contrast vividly in autumn with the brilliant yellow of the deciduous *Ginkgo biloba* planted among them. Araucarias grow particularly tall under the conditions here, and there are fine old examples of the bunya bunya, *Araucaria bidwillii*, the hoop pine, *A. cunninghamii*, and the Norfolk Island pine, *A. heterophylla*.

The dolerite soil of the Gardens makes it suitable for the growing of conifers. The climate, with a January mean maximum temperature of 22 degrees C dropping to a minimum of 11.4 degrees, and a July mean maximum of 12 degrees and minimum

Left: The Arthur Wall, separating the Gardens from the street, was built in 1829 and was heated.

Bottom left: The Eardley-Wilmot Wall, built by convicts between 1843 and 1846, originally defined another of the Gardens' boundaries.

Right: Superintendent Davidson's cottage was built in 1829.

of 4.4, is appropriately cool. The average rainfall is 622 mm, and watering of the gardens was a problem until the construction of the lily pond in 1848.

There is also a fine collection of temperate deciduous trees, and if you visit in the autumn when the snow swirls round the top of Mount Wellington, you will see large specimens of oak, beech, ash, birch, prunus, pistacia, dogwood and poplar colouring brilliantly.

As you look down the first hill, your eye goes past the rose garden, with its modern hybrid tea and floribunda roses, to the Eardley-Wilmot Wall, which separates the Botanical Gardens from the grounds of Government House. Built partly to provide work for idle convicts, it offered (together with the Arthur Wall) protection from the wind that sweeps up from the River Derwent. As Hobart is in the path of the Roaring Forties, wind was a trial to be reckoned with in the planning of the Gardens.

Most of the old part of the Gardens, planted in the gardenesque style, is sandwiched between the two protective walls, with the planting according to landscape possibilities rather than to a systematic scheme. Newer areas of planting are outside the Eardley-Wilmot Wall and stretch along the banks of

the Derwent. Within the walls, the oldest tree is a much-doctored English oak, *Quercus robur*, planted in the 1830s outside the sandstone cottage originally built for the first superintendent of the Gardens, William Davidson, in 1829.

William Davidson moved into the cottage early in the same year, to begin the gardens with the aid of ten convicts. He had taken the appointment at the age of twenty-four, having migrated to Launceston from his native Northumberland in 1827. He brought with him the reputation of having won many horticultural and botanical prizes in Durham and Newcastle-upon-Tyne, and his baggage included two thousand grape vine cuttings and eight hundred fruit trees. He stayed in Launceston for nine months, planting out his vines and trees at 'Marchington', the country home of a friend from Northumbria, before being offered the Government Gardens post. A year after his appointment he applied for a number of worked trees from George Town near Launceston for the new Gardens, and ordered from England a large number of trees and seeds.

When he took over the Gardens, however, they had already been established for ten years, probably just for the growing of fruit and vegetables. It is not

Top left: 'Jardin Botanique d'Hobart-Town', a lithograph by the French artist Le Breton taken from Dumont d'Urville's *Voyage des corvettes l'Astrolabe et la Zeleé 1837-1840. (W.L. Crowther Library, State Library of Tasmania)*

Left: The Botanical Gardens depicted by Marianne North in 1881. *(Royal Botanic Gardens, Kew)*

known what they were like at that time as the historical papers of Governor Sorell have been destroyed. In 1818 Sorell had taken possession of the 50 acres (20.2 ha) that were to become the Domain, the site of Government House and the Botanical Gardens from one R. W. Loane, who had no legal title to them. (The land had been part of Hangan's Farm, an original grant of land to John Hangan in 1806.) In the *Hobart Town Gazette and Southern Reporter* of January 1819 there is an entry relating to the salary of a J. Faber – £5 for the quarter-year for superintendence of the Government Garden and Grounds. The Gardens can thus probably be regarded as dating from 1818.

The first mention of botanical gardens was made by Governor Arthur in a dispatch of 1828: 'It was my wish that a Botanical Garden should be proceeded with in the Domain and I had hoped it may have commenced this season.' He added that nothing had been done 'in Collecting Plants, Shrubs, etc. with which this colony abounds. It is discred-

itable not to stir in this, and I am anxious about it as I find it remarked by strangers'.

Since 1826 Governor Arthur had had plans for rebuilding Government House on the new site of Pavilion Point next to the Government Farm, as the original Government House on Franklin Square in the centre of Hobart Town was, even as early as 1811, considered unsafe in a gale. Arthur's proposed 'palace' had progressed as far as having the sandstone blocks for it laid in place upon the ground. But the prohibitive cost of building had forced the plans to be abandoned. He was therefore keen to start a garden in both the vicinity of Government House and the Domain, as well as the Botanical Gardens themselves. And while plans for the building of Government House were remade, and the building completed by Sir William Denison at a cost of £120 000, the Gardens in the Domain evolved.

By 1830 the area enclosed for the Botanical Gardens was 13 acres (5.3 ha), and twelve gardeners and thirteen convicts were needed to keep them and the

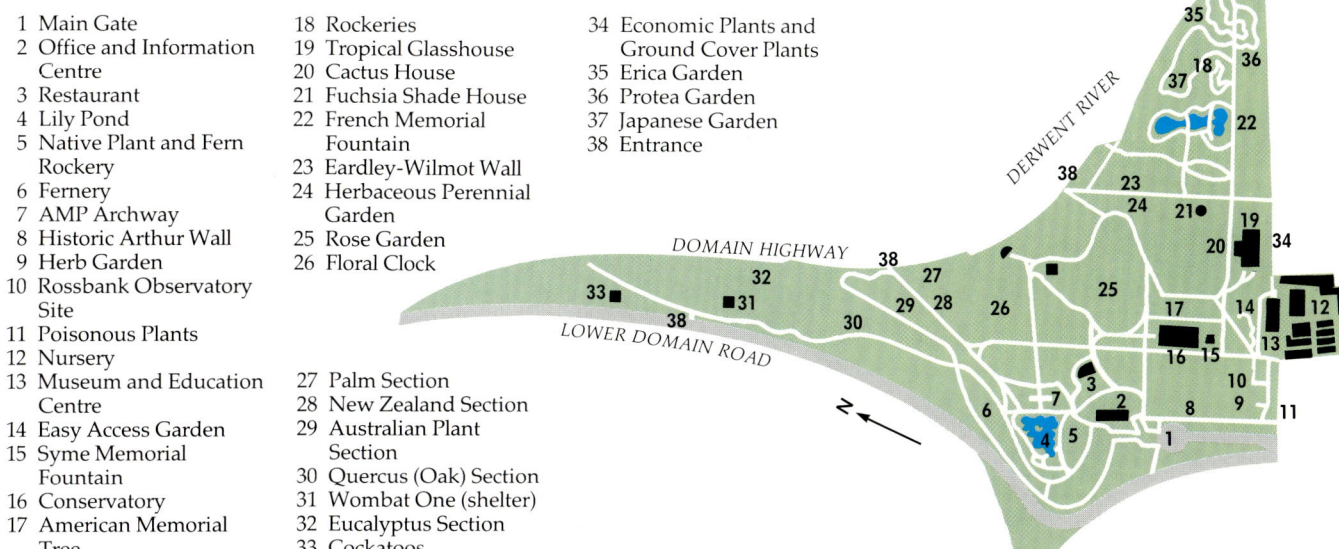

Domain in order. £300 was spent each year on their development, with £100 for the salary of William Davidson. They became such a popular resort that Governor Arthur was forced to close them on Sundays because the crowds were so unruly. A glasshouse was built to Davidson's design, and two hundred pineapple plants were installed there. In response to Governor Arthur's concern regarding the collection of native plants, Davidson explored Mount Wellington and collected 130 species. These flourished in the new Gardens, as did fruit trees. He was entrusted with a hive of the first European bees imported into Tasmania by Dr T. B. Wilson, and was so successful in nurturing them that Governor Arthur was able to send another hive to Governor Bourke in Sydney.

William Davidson resigned in 1834, and died in 1837. Martin Tobin was appointed superintendent to succeed him, and in 1840 a Mr Herbertson took over. The Gardens were supported with funds from the public purse, but in 1844 they came under the jurisdiction of the Royal Society of Van Diemen's Land for Horticulture, Botany and the Advancement of Science, the oldest scientific institution in the Empire outside of the British Isles. In 1848 it combined with the Tasmanian Society to become the Royal Society of Tasmania – a body that still exists.

Funds for the Gardens fluctuated wildly, as the Society supported most scientific organisations in the colony including those concerned with meteorology, astronomical observations, the supervision of fisheries and stocking the lakes with salmon, the organising of international exhibitions and Tasmanian exhibits in exhibitions around the world, and a museum. A grant of £400 in 1844 was supposed to be annual. In 1877 it had increased to £600, but in 1880 it was reduced to £450, and a year later increased again to £600. In 1845 Francis Newman was appointed superintendent at £80 per annum and he began a plant exchange with other states. Two hundred and fifty new species were introduced, including fifty indigenous Tasmanian plants arranged in a border

on the northern side of the Gardens, a bed of New Zealand plants, and one of medicinal plants. Generally, though, the Gardens under Francis Newman's guidance were planted in the gardenesque style that characterises them today, with pretty flowering trees, shrubs and fruits. In 1845 Newman produced a catalogue of plants in the Gardens, and in 1848 the lily pond was built as a storage dam. It was also used to acclimatise freshwater tench for stocking Tasmania's rivers. By 1856 public attendance at the Gardens had risen to 13 259 and the cultivated area had increased to 20 acres (8.1 ha), with many new grasses, grains and fruit trees being introduced to the colony through the Gardens.

A catalogue printed in 1857 lists, among two thousand plants, 105 varieties of apples being grown there, forty-four of pears, twelve of cherries, and numerous kinds of gooseberries and raspberries. By 1875 there were seven thousand different plants in the Gardens, many of which were fruit. 'Fruit grew better here than in England', noted Marianne North in her *A Vision of Eden*, a journal of her botanical travels from England to the southern Hemisphere in 1881 during which she painted scenes from the Hobart Gardens. Tasmania grew half the world's jam, she added. She deplored the lack of indigenous flora in the countryside, and the fact that, with its hawthorn hedges, it was all 'far too English'.

An earlier visitor, Frederick Mackie, writing in his Quaker Journals of 1857 (*Traveller Under Concern*) was also critical of the approach to native flora. Of his visit to the Botanical Gardens he wrote:

There are many fine specimens of native plants but few, if any, had their names attached. But those I was familiar with, being commonly cultivated in England were conspicuously named. This was vexing as I was desirous of becoming acquainted with the names of indigenous plants.

There was always immense botanical interest in Tasmanian native flora. An amateur botanist, Ronald Campbell Gunn, who migrated from Eng-

Left: View of the Derwent River from the Botanical Gardens.

Right: Grove of birch trees.

land in 1830 and became secretary to Governor Franklin, and later Commissioner for Crown Lands, explored the Tasmanian hinterland for many years. According to J. H. Maiden, director in 1896 of the Sydney Botanic Gardens, Gunn 'took Tasmanian botany from where Robert Brown had left it, to almost the present day'.

In 1840 Ronald Gunn met Joseph Hooker, son of the director of the Royal Botanic Gardens at Kew, Sir William Hooker. The younger Hooker had travelled to Tasmania as a ship's surgeon and spent a year in the wild, botanising. Wrote Hooker of R. C. Gunn, 'There are few species of Tasmanian plants he has not seen'. (Today over fifty species, including the hardwood *Eucalyptus gunnii*, bear his name.) From his travels with Gunn, Hooker in 1859 published his *Florae Tasmaniae* in two volumes, and his research led to his belief in the theory of descent by natural selection shown in plant geography, which became 'The greatest buttress to the theory of evolution', according to Darwin, whose *Origin of Species* was published in 1859. R. C. Gunn's collection of specimens was given to the Tasmanian Botanical Gardens for the herbarium, as were eighty-eight species of Asiatic plants collected by Joseph Hooker, who became a great plant-collector and eventually succeeded his father as director of Kew Gardens.

Most of R. C. Gunn's original herbarium material was sent to Sydney in 1930. The Tasmanian Herbarium had begun in 1845, when the secretary of the Royal Society, Dr George Story, began collecting dried specimens of local plants; some of Gunn's material and that of William Archer and W. A. Weymouth were added to it in 1913. In 1900 it became a branch of the Department of the Tasmanian Museum and Art Gallery and is now housed in the University of Tasmania's Department of Botany. In 1987 about 120 000 specimens, predominantly Tasmanian, are cared for at the herbarium by botanist Dr A. E. Orchard, and botanical research is done here.

Living plants continued to be collected at the Botanical Gardens during Francis Newman's superintendence from 1845 to 1859. The native plant collection was strengthened between 1859 and 1903 while Francis Abbott Jnr was the superintendent. He had been appointed as Newman's apprentice gardener, and was convinced that the role of a botanical garden should be 'educational, scientific and practical in its tendencies'. The possibility of establishing a class ground showing the order of plants was brought up at a meeting of the Royal Society in September 1881 following a member's visit to the Adelaide Garden of Dr Richard Schomburgk. The member, T. Stephens, claimed that

The importance of encouraging the study of botany by some practical teaching of this kind, a little of which will do more than a library of books will be readily admitted and a recognition of the fact that the Royal Society's Gardens are designed to afford some educational facilities will certainly not lessen their claim to support at the cost of the State.

The class ground was built.

Francis Abbott Jnr had been an apprentice of Francis Newman, and he took the view that a botanic garden was a scientific organisation, its role as a public park secondary to that. He stopped the propagation of fruit trees for the general public, although sales of surplus plants, seeds, cuttings and fruit trees at prices comparable with those charged by market gardeners and nurserymen continued. For the first time the general public was allowed free access to the Gardens every day of the week, and the sea wall along the beach frontage was built with an esplanade alongside it.

The importance of Tasmania as a summer resort was also becoming recognised, and though the fortunes of the Royal Society were at a record low in the early 1860s, it was thought inadvisable that the Gardens should be deprived of funds.

The Society has ever felt that, irrespective of the great scientific value it was a duty in reference to the more immediate interest of Hobart Town and the colony generally to keep the gardens in such order as should enhance the advantages of the place as a summer resort.

Nevertheless funds for the maintenance and improvement of the Tasmanian Botanical Gardens were far sparser than those provided for the Gardens of Sydney, Melbourne or Adelaide.

By 1885 the Botanical Gardens had been given back by the Royal Society to the Crown and were being administered by a body of trustees, six of whom were chosen by the Royal Society.This arrangement lasted until 1950, when an Act of parliament appointed a board of trustees comprising four government representatives, a representative of the University of Tasmania, one from the Hobart City Council and one from the Royal Society. They meet round the vast plank of Huon pine that is the committee table in the Gardens Museum, once Davidson's cottage, and which now displays samples of local timber, economic plants and home-spun wool dyed with Tasmanian plant extracts. In the past few years there has been an emphasis on planting Tasmanian flora gathered from the wild. There are 1200 native species in the Gardens from an estimated 10 000 species indigenous to Tasmania.

Living specimens of the famous Tasmanian rainforest trees are among the new additions to the Gardens. In a simulated rainforest terrain with a waterfall, and under a slatted timber roof, grow species of the Huon pine, *Dacrydium franklinii*; the King Billy pine, *Athrotaxis selaginoides*; the celery top pine, *Phyllocladus aspleniifolius*; the leatherwood tree, *Eucryphia lucida*, which produces Tasmania's famous honey; the Tasmanian laurel, *Anopterus glandulosus*; and many Tasmanian tree ferns from the western part of the island.

The policy of the Gardens today, explains Tony May, the director, is to grow as many varieties of plants as the weather and the soils will support. Many, such as the rainforest species, are grown with the help of artificial microclimates in special buildings. The most spectacular of these is the conservatory built in 1939 and housing four displays a year of potted plants. These Gardens are considered to have the best indoor flower displays to be seen in a botanic garden in the country, with ten thousand potted plants raised annually.

In the spring there are the calceolarias, schizanthus and *Primula obconica*. In summer there are tuberous begonias, hydrangeas, cymbidium orchids and lobelias (both in hanging baskets), impatiens and coleus. Autumn has bright chrysanthemums, tuberous begonias and coleus, and in winter there are the cyclamen, cinerarias and *Primula malacoides*. New varieties of seeds are obtained from various seed conferences around the world to help make the displays outstanding. They are arranged in tiers in the building, which has recessed bays for individual displays of flowers – enabling many different displays to be exhibited. The base is sandstone from the demolished Hobart General Hospital, the roof is of glass, and it was built through the enthusiasm of a member of parliament, the Hon. H. Shoobridge, a relative of the second director of the Australian National Botanic Gardens in Canberra. The superintendent of 1939, I. V. Thornicroft, designed the Conservatory. He was a horticulturist who had trained at Kew Gardens, the Dahlem Gardens, Berlin, and Edmonton University, Canada.

There is also a heated tropical house, which was built at about the same time, and in a temperature of 29.4 degrees C, with a humidity of 85 per cent, tropical plants are arranged in a rainforest setting, with tall *Ficus* species sheltering smaller palms and ferns. Fuchsias in this climate also need shelter, and in 1985 a modern slatted-wood fuchsia house was

Above: The Conservatory display in the Botanical Gardens is considered to be one of the finest in the Commonwealth.

Top right: Eucalyptus globulus, the Tasmanian blue gum, in the Botanical Gardens.

Right: This 19th century style summerhouse suits the garden design.

Left: Famous Tasmanian rainforest timber trees and ferns from the western part of the island grow in a shade house.

Bottom left: The shade house seen from the outside.

built to replace one designed in 1958. Today 250 varieties of fuchsias fill the shadehouse.

Outside in the elements, planting in the new sections is arranged according to ecological and taxonomic groupings rather than for landscape effect. This is evident in the new planting outside the Eardley-Wilmot Wall, which had arches cut into it to lead to an additional Gardens area of 2.2 hectares transferred from the Government House grounds in 1950. Here you come to the fountain of Huon pine designed by Tasmanian sculptor Stephen Walker in 1970 to commemorate the 200th anniversary of the French exploration of southern waters. Round it is a collection of alpine and rockery plants.

A collection of New Zealand native plants, looking healthy in the Tasmanian climate, includes specimens of *Hoheria, Pittosporum, Coprosma, Leptospermum* and *Hebe; Phormium* is seen along the banks of the River Derwent. In a sheltered hollow there is a collection of palms from all over the world, and along the river bank, colouring well in the autumn, is a large collection of oaks, the best, apart from that in Melbourne, to be seen in a state botanic garden.

A newer section of genus planting is that for *Eucalyptus.* The Tasmanian blue gum, *Eucalyptus globulus,* seen planted throughout the world, is not

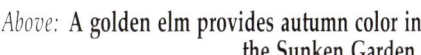

Above: A golden elm provides autumn color in the Sunken Garden.

Right: Newly constructed beds for the disabled provide easy viewing of plants.

Below: The French Memorial Fountain of Huon pine, designed by Stephen Walker in 1970, commemorates the 200th anniversary of French exploration in the South Pacific.

Autumn foliage and green lawns are loved features of the Gardens.

a native of the Hobart area, but one fine example can be seen from the entrance to the Gardens, at the bottom of the hill. Many other Tasmanian native eucalypts well known for their timber have been planted, as well as species from other states, in the new *Eucalyptus* section. An older section of native plants, including the indigenous Tasmanian shrubs *Helichrysum argophyllum* (the bright yellow, scented paper daisy), pimeleas, olearias, prostantheras and elaeocarpus, is in the part of the Gardens between the two great walls.

Alpines and native rock plants can be seen in the Oriental water garden round the lily pond in the old section of the Gardens. Here on the edge grows a large clump of *Gunnera*, the giant rhubarb-leaved Oriental water plant, and on the lake you can see indigenous waterlilies of red, white, yellow and orange on short stems. These are cold-water nymphaeas quite unlike the long-stemmed blue and pink species seen in tropical climates. The pond is shaped like a map of Tasmania and surrounded by terraced planting that includes many annuals.

On 17 September 1967 the prefix 'Royal', granted by Queen Elizabeth II, was added to the title of the Gardens. In keeping with their educational role, many new sections have been added recently, among them a Japanese garden, a garden for the disabled, and a large herb garden growing many varieties of culinary and medicinal herbs and those yielding dyes. There is a section on poisonous plants.

But the overall impression is still that of a landscape garden. The best time to visit the Gardens? According to *Walch's Tasmanian Guide Book of 1871*, 'In the first days of summer, when the roses have just burst into full beauty and the greenness of spring time yet lingers on the hill and lawn, these gardens are at their loveliest'.

4

THE ROYAL BOTANIC GARDENS, MELBOURNE

Of Australia's botanic gardens, the Royal Botanic Gardens, Melbourne, are considered to be the most beautiful.

'It is worth crossing half the world to see them', wrote Edward Hyams in *Great Botanical Gardens of the World*. 'In the matter of sheer beauty they are superlative.' The style was originally gardenesque, with plants gathered from all over the world growing in an idealised natural eighteenth-century landscape. Over the years the gardenesque style has blurred into the picturesque. The magic of the Garden, according to Hyams, lies in the 'very skilful disposition of lawns, the layout of paths, the use of water, and above all the placing of specimen trees and groups of trees [so that] it presents itself as a long series of landscape pictures – vistas, panoramas, prospects and perspectives, all of quite exceptional beauty'.

The position of Melbourne's Gardens, only 1.5 kilometres from the centre of the city and in the middle of a 250 hectare chain of parkland stretching along the banks of the Yarra River, is perhaps the most sylvan of Australia's botanic gardens, with the city skyscrapers barely visible. A 1980 ordinance restricting buildings in the vicinity to a height of no more than 12.19 metres helps to ensure their tranquillity, as do three of the boundaries: the Yarra River, the grounds of Government House, and the Domain. But though blending with the environment, the Gardens are most definitely divorced from it, and the task of the eighty or so staff employed by the Victorian Department of Conservation, Forests and Lands, which administers the Gardens today, is to maintain them to the design of 1908, while rationalising the botanical collections contained in them.

They are essentially the work of William Robert Guilfoyle, the fourth director of the Gardens, whose term of office from 1873 to 1909 was devoted to redesigning completely the Gardens created by the three previous directors, so that, of the original Gardens, only a bank of conifers on the western side remains.

To fashion his masterpiece, Guilfoyle used the beauty of the undulating site of 86½ acres (35 ha) sloping down to a large lagoon, with its opportunities to create microclimates for a wide spectrum of flora. Three things helped him: his natural ability in horticulture and landscaping gained from an apprenticeship with his father, Michael Guilfoyle, a nurseryman and garden-maker from Double Bay, Sydney; the climatic advantages of Melbourne, which are among the best in the world for horticulture, akin to those of California or Greece; and the best collection of flora of any botanic garden in Australia, gathered by his predecessor Ferdinand Mueller. His genius was in compiling the advantages into an original work of art.

The Gardens are based on nine main lawns placed horizontally along the site, semi-enclosed and interconnected, leading eventually to the 4.2 hectare lagoon at the bottom of the 'picture'. The lawns are linked by broad curving paths, with 'tunnels' of planting so that there is constant contrast between light and shade, wide and narrow, which heightens the interest as you are led to explore further. At every turn, the shapes of specimen trees take the eye – palms, conifers, willows . . .

Guilfoyle's sympathy for the harmonious position of trees in the landscape was based on a broad experience of southern hemisphere plants and their horticultural requirements, and thus he was able to use indigenous trees and shrubs to perform a similar artistic role to that of the great English landscapists with their cedars, beeches, chestnuts and oaks. Instead of cedars, a fine collection of araucarias shapes the landscape: the bunya bunya pine, *Araucaria bidwillii*, the hoop pine, *A. cunninghamii*, the Captain Cook pine, *A. columnaris*, the Norfolk Island pine, *A. heterophylla*, and the graceful pine, *A. rulei*, named after the Melbourne nurseryman J. J. Rule, whose collector found it near the summit of a volcanic island near New Caledonia. It is one of only two specimens in Victoria (the other is in the Geelong Botanic Gardens and came from the Melbourne parent). Other araucarias, *A. muelleri* and *A. hunsteinii*, complete the collection.

Overleaf: The lagoon of the Royal Botanic Gardens, Melbourne before the weed was removed. The lagoon is the central feature of the landscape design.

The most beautiful tree in the Gardens is considered to be the Murray River pine, *Callitris columellaris*, planted on the Hopetoun Lawn in 1860. It is an Australian native shaped rather like a Mediterranean cypress. Exotic conifers from warm climates grow exceptionally well. There is the 30 metre high, enormously wide Monterey cypress, *Cupressus macrocarpa*, probably planted in the 1860s, also on the Hopetoun Lawn, and the rare Chinese conifer *Keteleeria fortunei*, with a cedar-like habit.

Guilfoyle used palms and the shapes of the native grass tree, *Xanthorrhoea*, to create punctuations in open lawns with great artistic skill. His skill with foliage is seen at its liveliest in the rockeries, where bizarre succulents are entwined in rocks reminiscent of grottoes and with lava encrusted with shells; these rockeries are quite unlike the modern kinds, of which the garden designed by Ellis Stones with rocks and native plants near the Nymphaea Lake is a good naturalistic example. The buildings – summerhouses and pavilions – are also attractive, with no two styles the same. Within half an hour's walk of each other you can come across the Swiss-style William Tell chalet on the edge of the lake and the Guilfoyle-designed 'Temple of the Winds' on the bluff above Alexandra Avenue. (Instead of the traditional acanthus on its Corinthian columns, this 'temple' has staghorns from the rainforests of northern New South Wales.)

Though the Gardens make up an integrated landscape, some eleven thousand labelled botanical species (excluding the glasshouse collections), contribute to the scene. Melbourne's climate, with a January average maximum temperature of 26.5 degrees C and a minimum of 14.9 degrees, a July maximum of 13.5 degrees and a minimum of 6.2 degrees, an average annual rainfall of 660 mm and no frosts, will support most flora. Only the humidity-loving cold-climate larches and firs fail to thrive.

The Gardens have the finest collections of plants from southern China in Australia, and the collection of natives from New Guinea, often valuable for experimental horticultural and economic use, is also the most comprehensive grown here. The viburnum collection, with forty species, is reputedly Australia's finest, and the collection of approximately 180 camellia varieties is recognised as being excellent. The collection of forty oak varieties is Australia's largest.

Though attractive in the autumn when the deciduous trees have turned colour, the Gardens are considered to be best in spring when the new growth on the huge trees of the oak lawn – the white oak, *Quercus alba*, the scarlet oak, *Q. coccinea* and the daimyo oak, *Q. dentata* – is brilliant. At this time the coloured tunnels of azaleas and rhododendrons that link the lawns are flowering, and the flowering trees silky oak, *Grevillea robusta*, cape chestnut, *Calodendrum capense*, and the many jacarandas in the Gardens soon follow.

The plan to have a botanical garden for the city of Melbourne is believed to date from 1836 – a year or two after the founding of the city – when Charles La Trobe, the superintendent of the Port Phillip District of the Colony of New South Wales, entered an expenditure for gardens in his report to the New South Wales Government. The Gardens on their present site were not begun until 1846, however. It was the second choice for a site. The first, 49 acres (19.8 ha) of Batmans Hill overlooking the town of six thousand citizens, had been selected in 1841 before an application by Melbourne for £750 to run them had been approved by the New South Wales Government. The money was refused, and by the time it was eventually given in 1845–46 the site had been abandoned owing to the rapid growth of the city. (It was at what is now the western end of Collins Street, and the hill was removed to make way for the Spencer Street railway station.)

With the money in hand, the need for a new site became urgent and Charles La Trobe, as chairman of a small committee of leading citizens, chose 'a beautiful small valley to the south side of the Yarra River half a mile east of the town'. La Trobe, an amateur botanist, geologist, and hunter of butterflies and beetles, described it as infinitely superior to the previously spoken of site, 'a veritable garden of Eden' in fact. It was also too steep to be used for city buildings. On its undulating slopes, leading down to a string of marshes, were an Aboriginal mission and a dairy, and it was wooded with banksias, wattles, and a thick spread of river red gum, *Eucalyptus camaldulensis*.

There are still two river red gums today in the Gardens and one, heavily doctored but surviving well, is the 'Separation Tree', around which celebrations for the separation of Victoria from the state of New South Wales, which took place on 1 July 1851, were held.

With his site agreed on, La Trobe set about finding an administrator for the Gardens. He chose John Arthur, a Scot who had worked as a botanist in Perthshire before migrating to become head gardener for a large estate and to establish his own nursery in Melbourne. Arthur was engaged on 1 March 1846, and on 16 March Governor Gipps of New South Wales approved the foundation of the Botanic Gardens in Melbourne.

'Arthur's Elms', *Ulmus procera*, the four magnificent old trees on the Tennyson Lawn of the Gardens, were among the first exotic trees planted – and they were brought from Arthur's own nurseries. Lawns were also among the first plantings to be established, as La Trobe was anxious to hold garden parties on them to show that at last something was

Top right: One of the many fine old Australian trees, a *Melaleuca* species, thrives in the Gardens.

Right: The Separation Tree, a *Eucalyptus camaldulensis*, round which the civic celebrations for the separation of the State of Victoria from New South Wales took place on 1 July 1851.

Far right: *Eucalyptus camaldulensis*, the river red gum, is one of the few indigenous specimens extant in the Royal Botanic Gardens, Melbourne.

being done. The 5 acre (2 ha) site adjoining what is now Anderson Street on the north-east boundary running down to the lagoon was fenced and chosen as the most suitable place to start, and Arthur made the first lagoon walk round the historic gum trees. He followed the designs of Henry Ginn, the colonial architect, and though there are no plans extant, it is certain they typified those of grand English landscape gardens. After three years in office Arthur died, supposedly after drinking water from the often cholera-contaminated Yarra River. La Trobe was reported to be extremely upset, for the two men had been friends, and he chose another Scot, John Dallachy, to replace Arthur.

Dallachy, although he had landscape experience at Haddow House, the seat of the Earl of Aberdeen, was primarily a plantsman. In Scotland he had cultivated the plants of New Holland and later worked at the Royal Botanic Gardens, Kew, with Sir William Hooker before migrating to Melbourne to become head gardener for the 6 acre (2½ ha), Italianate/English landscape garden of J. B. Were.

During his period as superintendent of the Gardens from 1849 to 1857, he increased the number of species to six thousand – one thousand of them native plants, many of which he had collected during his trips into the hinterland of the state. He began a systems garden, along the lines laid down in John

Lindley's *Vegetable Kingdom* (1846), and plants were labelled with their scientific and popular names on iron labels brought out as ships ballast; many of them are still there today. He also planted specimen trees, and his *Araucaria cunninghamuii, A. heterophylla* and *Sequoia sempervirens* are still there today on the Eastern Lawn. Other plantings by Dallachy can be seen on the south-eastern slopes of the Gardens.

Dallachy worked initially under great duress: there were no nursery buildings, he had to make his own huts of brush, and manpower was short due to the exodus from the city brought about by the Victorian and New South Wales gold rushes. He slowly proceeded with planting and by 1851, in time for the Separation celebrations in the Gardens, the area under cultivation was 6 acres (2.4 ha). Underground drains were laid and a glasshouse was built.

Over the years Dallachy's field-collecting of living plants absorbed more and more of his time. Dried specimens were also needed for the foundation of a herbarium collection. He was stimulated in collecting by the material being sent to the Gardens from Adelaide by Ferdinand Mueller, a trained chemist and graduate in botany from Kiel University in Germany, whose passion was botany. Mueller's

interest in the Melbourne Gardens, and in supplying them with seeds, increased so much that in 1852 he was appointed by La Trobe as the first government botanist for the Botanic Gardens in Melbourne. Because of increasing links with the Royal Botanic Gardens at Kew, through seed and plant exchange and herbarium material, the post of government botanist had become essential.

Ferdinand Mueller lived in a newly built cottage at the Gardens and, with John Dallachy becoming increasingly immersed in field trips and plant-collecting, sometimes as far away as Rockingham Bay, Queensland, he became increasingly involved with administration. Finally, in 1857, he superseded Dallachy as director. Dallachy later became the official curator and continued working on and collecting plants for the Gardens.

Mueller had already recommended, as a member of the Botanic Gardens Committee which administered the Gardens, the construction of 'Gardens House', a single-storey stone building which, with the addition of a second storey in 1861, became both the office and the residence of the director.

During his administration, which lasted until 1873, he increased the planted section of the Gardens to 66 acres (26.7 ha), with 338 acres (136.8 ha) of the adjoining Gardens and reserves being cared for. He also developed the Gardens as a scientific and educational institution. Addressing an industrial group in Melbourne in 1853 on the 'Objects of a Botanic Garden in relation to Industry', he said:

The objects must be mainly scientific and predominantly instructive. As a universal rule it is primarily the aim of such an institution to bring together with its available means the greatest number of select plants from all the different parts of the globe; to arrange them in their impressive living forms for systematic, geographic, medical, technical or economic information and to render them accessible for original observations and careful records.

He introduced a systems garden that took up about 3 acres (1.2 ha), with 1700 plants from a hundred genera. In 1857 he was also given temporary control of the collections of the Zoological Society of Victoria, just across the Yarra from the Botanic Gardens. Gradually he established aviaries and animal enclosures in what is now Fern Gully.

Mueller planned to introduce and acclimatise songbirds from overseas, and a consignment of nightingales was released in 1857. An aviary was placed in dense shrubbery between a bridge and barren area facing the lagoon, a forerunner of the 'walk-through' aviaries of a century later. In the 1860s caged birds were introduced into the Gardens; pairs of thrushes, blackbirds, starlings and skylarks were released and began to breed in 1861. Llamas for alpaca and angora goats for their wool were added.

In 1858 he began a pinetum on the western part of the lagoon, planting Aleppo pines, Norfolk Island araucarias, and groups of other conifers. Walks and straight paths in the Gardens were lined with avenues of trees, mainly the native silky oak, *Grevillea robusta*, the Illawarra flame, *Brachychiton acerifolius*, and the blue gum, *Eucalyptus globulus*, of which he was so fond that his nickname became Blue Gum Mueller. His fondness for conifers produced by 1865 ten thousand pines in the Gardens, the Government House reserve and the Domain reserve. He established avenues of forty-two species of trees. By 1867 there were 21 000 pines planted in these reserves. Labelled plants in the Gardens reached two thousand species. Mueller threw himself into the marriage of zoological specimens with botanical ones, and in 1859 the name of the Gardens was changed to Botanical and Zoological Garden.

During his directorship Mueller arranged for the distribution of plants and seeds from the Gardens to centres throughout Victoria, 36 000 in the 1860s. He provided the planting nucleus for the botanical gardens of towns including Ballarat, Bendigo, Castlemaine and Kyneton. Many native plants had come from his field trips round Victoria, the Snowy Mountains area, the Northern Territory and northern Queensland. Eucalyptus seed he sent abroad established planting that drained Italy's Pontine marshes, provided shade in Abyssinia and timber for Transylvania. Australian flora was dispatched

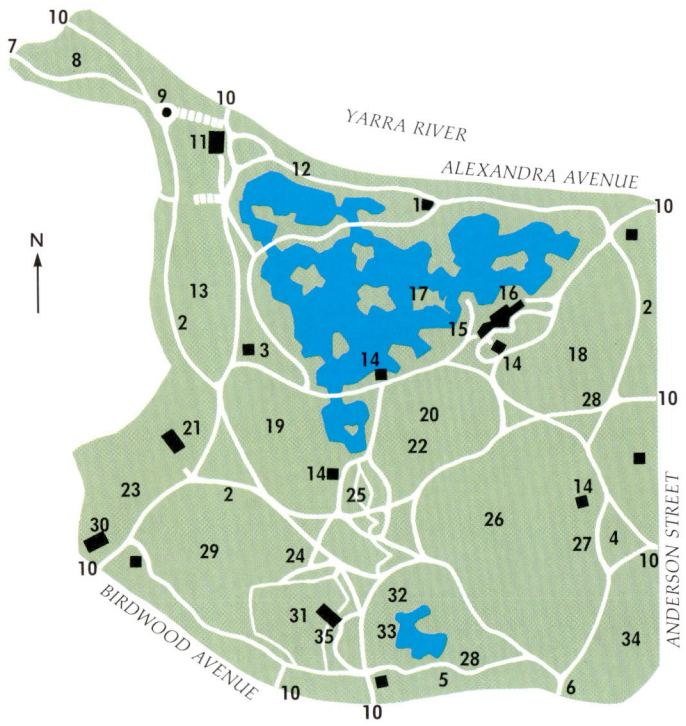

from the Melbourne Gardens to most corners of the world. Mueller belonged to three hundred international scientific organisations and wrote about three thousand letters to colleagues every year.

The herbarium, begun with Dallachy's collections, was now housed in a new building in the Domain, west of the western entrance to the gardens and with space for 160 000 specimens in 1861; by 1865 it had 286 000 specimens. Even today Mueller's herbarium probably represents half of Australia's collections. He arranged the preparation of thirty-two garden lots on the western bank of the lagoon for trials of economically useful plants: cotton, ginger, coffee and tea. He experimented with grasses, and although the Gardens did not contain much lawn at the time, the mixture of buffalo and couch he developed as a hard-wearing strain to withstand foot traffic has been the basis for all the lawns of the Gardens since then. Mueller organised a reticulated water supply, connected in 1864, and con-

Right: **The Temple of the Winds, designed by William Guilfoyle in honour of Charles Joseph La Trobe, who chose the Gardens site.**

Centre right: **An Indian-style pavilion is typical of the eclectic style of summerhouses in the Gardens.**

Far right: **The William Tell chalet in a gothic rustic style.**

structed a tank for the growing of the *Victoria amazonica* waterlily – Australia's first specimen.

In 1859 the government was asked that Mueller should collaborate with George Bentham, the president of the Linnaean Society in London, in the production of a 'universal work on the Australian flora'. The seventh and final volume of *Flora Australiensis* appeared in 1878, and herbarium material collected by Mueller was sent to England.

As well as being decorated for his scientific work by most European countries, Mueller was made an honorary baron by the King of Württemberg in 1869. Later that year a forester, William Ferguson, was appointed as curator of the Botanic Gardens and inspector of state forests. He was to manage the landscaping and practical work at the Gardens, for although Mueller had built a garden that had become world-famous botanically, 'it did not', according to one contemporary writer, 'feed the hunger of the eye'. The Melbourne public was determined to have one that did, and which would be more of a pleasure park than the rigid rows of specimen plants they had.

The Gardens were under the jurisdiction of the Botanic Gardens Committee, which had direct access to the superintendent of the Port Phillip District. Mueller worked for a separate government department which was under the control of the chief secretary. (It later became the Board of Land and Works.) By the mid-1860s there was criticism of Mueller's directorship of the Gardens by the government and the general public. In 1868 he was asked to give a second report on his directorship (after his annual report), in which he justified his scientific achievements – the flowering of many rare plants for the first time, his success in reticulating

water. He countered criticism that he had introduced many weeds into the Gardens on plant-collecting excursions. This was all to no avail, for in November 1870 disputes between Ferguson and Mueller reached the notice of parliament and on 12 December parliament approved the sitting of a Board of Enquiry into the administration of the Botanic Gardens. The Board of four influential Melbourne citizens met twenty-two times, and only once was Mueller asked to defend himself. When the Board's recommendation was handed down, it constituted an attack on both Mueller's work and his style of management of the Gardens. They were to be not only botanic, but also horticultural in their character, said the Board, and it carried the recommendation that Baron Ferdinand von Mueller be in charge of the scientific branch, the laboratory, the class ground, and correspondence local and foreign, but for the practical management of the Gardens and reserves there should be a curator, a practical landscape and flower gardener, responsible for their proper design and maintenance. In July 1873 he was dismissed as director of the Gardens, though retained as government botanist. He took a room in Martins Hotel, South Yarra, and did not live inside the Gardens again.

As early as 1865 Mueller had been aware of the threat to his concept of the Melbourne Botanic Gardens, and had written to Joseph Hooker, who had succeeded his father Sir William as Director of the Royal Botanic Gardens, Kew. Hooker was a personal friend of Mueller, who had looked after Hooker's son Brian on a botanising trip round Australia and New Zealand. Hooker wrote advising him to:

Let the matter rest for a few months making no plaint nor stir and when public attention is withdrawn from your

present position, quietly consult your many powerful friends about your future career with a view to getting a thoroughly good artistic horticultural and decorative gardener to whom I would delegate the ornamental and practical gardening.

But the control of events was by 1873 out of Mueller's hands. His awkward personality and, it was said, the jealousy of the nursery trade whose influence he had usurped through his policy of distributing government plants throughout the state, conspired against him. But when William Guilfoyle was appointed as the new director at a salary of £500 a year on 31 July 1873, Mueller appears to have felt no personal animosity. He wrote to Guilfoyle, 'I should be glad my dear friend if you will first of all look for me when you arrive'. Mueller had proposed Guilfoyle as a member of the Linnaean Society, and had known and corresponded with the Guilfoyle family. He admired William's talent.

Guilfoyle had inherited a plan prepared by an amateur, Joseph J. Sayce of Caulfield, in response to a government-sponsored competition for a plan to beautify the south side of the River Yarra and the whole park area, which included the Domain and Government House. Sayce's plan comprised a pattern of wide curved paths, a large landscaped lake with islands and promontories, a fern gully, several extensive lawns, and a series of summerhouses. Guilfoyle approved of this plan in general but modified it.

First there was an initial period of cleaning up. Guilfoyle extended the space in the ferntree gully by moving all the birds and animals out to the Acclimatisation Society's headquarters at Royal Park. He complained that some of the small shrubs, gardenias, cytisus, justicias and palms were hidden under useless indigenous scrub, mainly acacias and leptospermums. He started to thin out plants and relocate them in order to produce more of a show. Seeds of colourful flowers were sown in new beds and he began to make amends for the almost complete absence of the extremely popular azaleas, camellias and rhododendrons. There was also an absence of lawn in the Gardens.

Guilfoyle began redesigning the area round the curator's office, planting camellias and azaleas and rhododendrons in the now famous 'tunnels'. By 1895 he wrote: 'this section is now complete, showing broad lawns, thickly clothed with verdant grass interspersed with various picturesque groups, single specimens and clumps of trees and beds of brilliant flowers . . .'. This, as a finished portion of the design, full of plants that had come to maturity, attracted great attention from the public. Early on, Guilfoyle obliterated all narrow paths and replaced them with sweeping paths 15 feet (4.6 m) wide, which followed the contours of the land. He wrote at the time: 'most people will agree with me that it is far easier to lay out a new garden than to remodel one in which blunders have been made . . .'

His replanting between 1875 and 1877 of existing flora, an operation that included moving some 2300 trees (some of them up to 10 metres in height), was fundamental to the design; he made world horticultural history by proving that it was possible to move large trees safely without a check in growth. These included araucarias, oaks and *Ficus*, planted two decades before by Ferdinand Mueller. Generally he provided large, mostly evergreen trees, central to each section of the Gardens, then graded down

Left: A tall hedge of lilly-pilly, *Acmena smithii*, is one of the distinctive features of the Gardens.

Right: Secluded corner with a foreground of *Phormium*, flax, in the bed of New Zealand native plants.

Bottom right: Palms and ferns from rainforest areas predominate in the Fern Gully.

the planting in size and colour to the edges of the groups.

William Guilfoyle's horticultural, botanical and landscaping experience had begun early in his life. Besides meeting most of the eminent natural scientists in Sydney at his father's house, he had studied botany with William Woolls, a botanist and classics master at Sydney College who had begun his own private school at Parramatta. Later joining his father Michael Guilfoyle's nursery firm, he became a partner and helped supply garden layouts and plants to clients.

Michael Guilfoyle had worked in London at the Exotic Nursery and had done landscaping with Joseph Paxton, the great Victorian landscape designer, before migrating and starting the nursery in Double Bay. His first large landscaping project was for Thomas Mort at 'Greenoaks', Darling Point – now the Bishopscourt and regarded as one of the finest gardens in Sydney. He also worked on the Sydney Botanic Gardens and the Domain, and designed layouts for and supplied plants to many clients in the Sydney and country areas.

Another influence on William Guilfoyle was his joining, as botanist, HMS *Challenger*, for a five-month expedition to the South Seas in 1863. During this

Above: A splendid *Xanthorrhoea australis*, the black boy, on one of the Gardens' lawns.

Below: *Xanthorrhoea australis*, black boys, mix felicitously with exotic echiums.

Right: Willows on an island in the lagoon give an Oriental tranquility to the water garden.

voyage he collected plants for the director of the Sydney Botanic Gardens, Charles Moore, including thirty-five varieties of dracaena, plus crotons, palms and poinciana trees. He observed tropical landscapes throughout the journey from Samoa, through the New Hebrides and the French territories, to Fiji.

Guilfoyle then helped his father in a cane-growing project on a 600 acre (242.8 ha) property Michael Guilfoyle had established at Cudgen in the Tweed District of northern New South Wales, growing crops of tobacco, cotton, coffee and peanuts. From here he was appointed to his position at the Melbourne Botanic Gardens.

His experience with the tropical landscape in both the South Seas and near Cudgen is obvious in the eventual design of the Gardens. Later he wrote: 'I think it should always be the aim of those in charge of public gardens to bring before the public scenes of such beauty not to be found elsewhere, by representing plants of a different character to those more or less common to the locality.'

Having first established the form of the planting, Guilfoyle's main preoccupation was with the shaping of the lagoon. When he took over, it was not a stable piece of water for it was connected with the River Yarra and subject to flooding and variation, and the banks were marshy and undefined. In the late 1870s he proposed to break the straight lines of the edges by constructing several promontories, as suggested by the plan of Joseph Sayce.

Guilfoyle's work on the lagoon was given an impetus in 1896, with the passing through parliament of the Yarra Improvement Act to stop the river flooding. This involved straightening the river, and subsequently isolating the lagoon, which became a self-contained body of water supplied by the runoff of rainwater from the area surrounding it. This added a further 3½ acres (1½ ha) to the northern front of the Gardens and provided a broad link between the east and west sections. (Guilfoyle had already added, in 1875, another 3 acres (1.2 ha) to the 45 acre (18 ha) garden by acquiring the cow paddock from the Government House grounds.)

Once the worry of flooding had been eliminated, Guilfoyle was free to make the lagoon the visual centre of the Gardens. He enlarged the larger islands, planting them 'boldly', and created a new one, Long Island, which was joined to the mainland by two bridges. With the extra land, he had a new thoroughfare built on the northern end of the Gardens, Alexandra Avenue, and on the 50 foot (15.2 m) escarpment overlooking it, the city and the newly formed lake, he began construction of the 'Temple of the Winds', modelled on the many examples from antiquity he had seen in travelling to England and Europe in 1888–89. He dedicated it to Governor La Trobe. He also built a palm house and glasshouses, dismantled Mueller's class ground, and built a large rosery on its site.

Of the botanic work of the Gardens Guilfoyle wrote: 'No necessity exists for allowing botanical correctness and landscape effect to clash in the de-

velopment of the Melbourne Botanic Gardens. To combine the two has been my design from the beginning.' The overseas experience gained from visiting the botanic gardens and landscape gardens of Italy, France and Britain (Scotland and Ireland) reaffirmed his conviction that the site for the Melbourne Gardens, with its undulating slope providing microclimates and a variety of aspects, was blessed. Only that of the Royal Botanic Garden, Edinburgh, rivalled it.

When Guilfoyle retired in 1909, after thirty-four years of reconstructing the Gardens, he knew he would not live to see their finished form. It took another fifty years for them to reach maturity.

Successive directors have modified Guilfoyle's planting slightly, and the bright beds of annuals have largely given way to perennial shrubs and smaller plants. The trees are today still planted as single specimens rather than in rows or groups, according to Guilfoyle's theory that a tree should take its natural shape in the landscape and that when rows of them are planted there is a chilling effect, and the sameness of foliage is accentuated. The subtleties of foliage and colour are still paramount in the planting.

The only new piece of landscape design within the Guilfoyle gardens is the rock garden designed by Ellis Stones, a Melbourne landscape architect renowned for his work in garden construction and with native plants, on the south-east end of the Nymphaea Lily Lake. This was built in the 1970s to exhibit Australian native rock plants, with money provided by the Maud M. Gibson Gardens Trust.

Left: Grotto of simulated molten lava and plantings of bizarre foliage in Guilfoyle's rockery.

Right: Agave americana 'Marginata' in the rockery.

Far right: Yuccas in the rockery show Guilfoyle's skill with sympathetic planting.

Bottom right: The Rock Garden by Ellis Stones, built in the 1970s on the edge of the Nymphaea Lake.

(The Maud M. Gibson Gardens Trust, established by a Melbourne-born benefactor in 1945 in memory of her father, William Gibson, a partner in the Melbourne company Foy and Gibson Ltd, has in two benefactions of 1945 and 1965 made possible various projects including the publication of scientific work and of books on the Gardens, as well as the construction of the Ellis Stones Rock Garden.)

Though the planting in the Gardens is not arranged in any systematic order but according to landscape needs, there are small areas of specialised planting, like the Ellis Stones rockery, the new Herb Garden, the rose garden, and an area comprising only New Zealand native plants.

It became apparent that there was a need for a more strictly botanical section of Australian plants than is contained in the native plants border along the southern boundary in the Gardens, and with money from the Maud M. Gibson Gardens Trust, a 320 hectare stretch of sand-dunes at Cranbourne, near the coast of Port Phillip Bay, was eventually purchased for an annexe to the Gardens. Though lacking topsoil (the site was used formerly for sand-mining) the Gardens here are planted with a 30 hectare perimeter of native flora and collections of banksia and grevilleas have been established. The annexe contains native heathland and woodland, pasture, and an area disturbed by sand extraction. The area of native vegetation will be retained, a native botanic garden will be planted in the disturbed area, and the pasture will be converted to an arboretum of Australian native trees.

Botanists and horticulturists working on the Cranbourne Annexe have their headquarters in the National Herbarium building in Birdwood Avenue, South Yarra. Here vegetation-mapping for the state, identification of plant species, conservation and preservation of endangered species and similar work is done. The building, donated by Sir Macpherson Robertson, a leading Melbourne citizen, was completed in 1934.

In the herbarium and library, the nucleus of the collections is based on those begun by Dallachy and enriched by Baron Sir Ferdinand von Mueller. A prized exhibit is an original copy of Bentham's *Flora Australiensis*, which could not have been compiled without Mueller's help. The herbarium now contains over a million plant specimens.

In the Gardens, though the basic design of William Guilfoyle is still adhered to, changes have occurred recently. The lagoons and lakes, originally planted by Guilfoyle with water species including a beautiful lily called *Nuphar luteum*, have been cleared of all plants. The *Nuphar* had become, by 1983, so rampant a weed that clear water space for reflections and for water birds was fast diminishing. The lake was desilted and dredged, thus starving the weed. Eighty thousand cubic metres of silt and mud were carted away.

The Herb Garden has recently been redesigned and replanted. The early garden included medicinal plants which, according to Guilfoyle's *Handbook or Descriptive Guide to the Gardens* of 1908, were grown in eight beds of nearly 500 plants in a special experimental section. Although it had been an important part of Guilfoyle's Gardens, featuring not only medicinal but also dye, fibre and forage plants, it had dwindled away and in 1982 a project to revive it was formed by the Gardens staff, the Friends of

Above: *Ricinocarpus pinifolius*, the wedding bush plant, is indigenous to the Cranbourne annexe of the Royal Melbourne Botanic Gardens.

Above: *Eucalyptus megacornuta*, the warty yate, grows at Cranbourne.

Below: General view of the sand dunes and swamps of the Cranbourne annexe.

Above: *Calothalmnus villosus* at the Cranbourne annexe.

Right: A tunnel of rhododendrons and azaleas in bloom in the Royal Botanic Gardens, Melbourne.

the Botanic Gardens and the Herb Society of Victoria. It has been rebuilt on the original site with the slope levelled and drained, a watering system installed, and the garden formed in a cartwheel-shaped pattern of brickwork beds. There are seven beds, each representing a botanical family or groups of families. Gate F in Birdwood Avenue is the nearest entrance to the Herb Garden. On the Princess Lawn there is a new herbaceous border featuring over 200 perennial flowering plants of original species which came from all over the world.

One of the most difficult tasks facing the staff at the Royal Botanic Gardens of Melbourne remains that of conforming to modern educational, horticultural and botanical needs without disturbing Guilfoyle's nineteenth-century design. Many of the original trees are dying or having to be removed for safety's safe. Replacing them with fully grown similar species is difficult and expensive. To maintain the Gardens like a set piece of art is a problem that the staff of no other Australian botanic garden has to cope with. There is criticism that this design stricture has caused a deterioration of 'botanical' as opposed to horticultural standards: Mueller, it is sometimes said, would have given Melbourne better 'botanical gardens'. The pursuit of Guilfoyle's beautiful vision is thought to be worth the effort, however. The English writer Sir Arthur Conan Doyle wrote of the Gardens: 'I spend such time as I have in the Melbourne Botanic Gardens, which is, I think, absolutely the most beautiful place that I have ever seen.'

5

SOME REGIONAL GARDENS IN VICTORIA

GEELONG BOTANIC GARDENS

At the entrance to the Geelong Botanic Gardens in Eastern Park, just out of the city, you look down the beds of bright annuals to the pompom fronds of a *Jubaea chilensis* palm – the eyecatcher at the end of the vista. It is a spectacular sight, and you are aware at once that these Gardens have a strong botanical bias.

The *Jubaea*, the Chilean wine palm, is one of only ten specimens in Victoria and is classified today as an endangered species in its native Chile. With another twenty-four trees in the Gardens listed on the National Trust's Victorian register of the state's significant trees, its presence in the Gardens underlines their scientific importance. Many of these trees were probably planted by the founding curator, Daniel Bunce, who is believed to have been trained at the Royal Botanic Gardens, Kew, in the late 1820s (the records of this period in the history of Kew Gardens are incomplete). Bunce began work at the Gardens in 1857, and in nine years, according to a report in the *Geelong Advertiser* of 1866, transformed them from a bare paddock to a state where 'There is no spot south of the line to compare with Mr Bunce's creation as a reflex of what English landscape gardening is capable'. In landscaping they outshone the Melbourne Botanic Gardens, which, in the 1860s, under the control of Ferdinand Mueller, were noted for their formality. The Melbourne Gardens were, however, famous for their scientific excellence, and Geelong kept pace with them in botanical innovation.

The Geelong Botanic Gardens then comprised 200 acres (80.9 ha) of parkland planted in a gardenesque manner on an undulating site and intersected by broad – 20 foot (6.1 m) wide – curving carriageways. In the middle of it all was an oblong section of 5¼ acres (2¼ ha), the botanical hub of the Gardens, and here were planted most of the rare native and exotic trees, some of which had come from Mueller as overflow specimens from the Melbourne Botanic Gardens and some of which were ordered by Daniel Bunce directly from his botanist friends in England and Europe. Shrubberies set along the land from north to south held both very rare plants and pretty favourites. Paths glistened white with a layer of crushed, raked seashells from the nearby Corio Bay. Ribbon planting of the most beautiful and newest flowers made brilliant beds in green lawns, and in the conservatories bloomed the most fashionable and latest tropical and temperate imported plants, including a 'magnificent specimen in flower of *Impatiens hookeriana*'. (Melbourne's Gardens had a similar specimen which had not bloomed.)

The plans of the 'nursery' section of the Gardens were once held in the Geelong Crown Lands vault and later were destroyed by floods, but some idea of the planting of the first years recently came to light when the British Library in London returned to Geelong a copy of Daniel Bunce's first catalogue of plants collated in 1859 and listing 2325 species in the Gardens.

Besides the fine landscaping of Daniel Bunce, the early Geelong Botanic Gardens had, by the 1880s, a fernery designed by Bunce's successor, John Raddenberry, the equal of which, according to a newspaper report of the time, did not exist in the colonies. In size it was slightly less than the mighty and beautiful Palm House designed by Richard Turner and William Hooker at the Royal Botanic Gardens, Kew, with Geelong's fernery 300 feet (91.4 m) long compared with Kew's 360 feet (109.7 m), and a height of 60 feet (18.3 m) instead of Kew's 66 feet (20.1 m). Among hundreds of ferns and palms, it enclosed the *Jubaea chilensis* and various other palms that are in the centre of the Gardens today. Unlike the Kew Palm House, Geelong's fernery was made of Australian hardwood, and by 1959 it had deteriorated so markedly that it was removed. But the shape of the fernery had an influence on the Gardens, as did the original layout by Daniel Bunce.

Today the Geelong Botanic Gardens are situated in the original botanical nursery oblong, set in the middle of curving avenues lined with *Cupressus macrocarpa* 'Lambertiana', *Pinus radiata*, *P. canariensis*, and the green lawns of the parkland of Eastern Park, the original Botanic Gardens. Along the eastern perimeter of the Gardens large conifers, mostly

Left: Old fashioned perennial plants echium and acanthus spill over gravel paths in the shrubbery of the Geelong Botanic Gardens.

Sequoiadendron giganteum, form a boundary, and on the western side the visual boundary is the longest avenue of bunya bunya pines, *Araucaria bidwillii*, in Victoria. All the trees were planted by Daniel Bunce. Paths are straight and form a grid on either side of the central flat section where once the fernery stood, and today they are covered with gravel instead of crushed seashells. On the western side the land slopes up from the centre quite steeply and is terraced, with the paths tiered to enhance the terrace effect. The shrubberies are full of similar planting to that of the 1860s, with some sixteen varieties of lilac, and many viburnums and bulbs. Overtopping them are some magnificent trees, including a fine purple beech, *Fagus sylvatica* 'Purpurea'. Clipped box hedges, about fifty years old, contain the shrubberies and add tension to the sprawling informality of planting within the beds. In spring the blue of echiums and lilacs and the pinks of prunus blossom spill from the hedging.

The flat central section still follows the pathways laid down within the fernery. It contains bright beds of annuals and a cactus garden, and you are led to the slatted wooden summerhouse of the 1880s, built by John Raddenberry. Beyond the summerhouse is an informal woodland garden of camellias and azaleas sandwiched between the bunya bunya avenue and a newly planted hedge of lilly-pilly, *Acmena smithii*, which has become the new boundary between the park and the Botanic Gardens. This woodland garden, 15 metres wide, was built in 1980 as an addition to the original garden, which ended

Above: Main axis of the Gardens leads to a fountain installed in 1912. The *Jubaea chilensis* palm, the Chilean wine palm, was planted in the 1860s.

Below: Ranunculus asiaticus forms an avenue of the main axis of the Geelong Botanic Gardens.

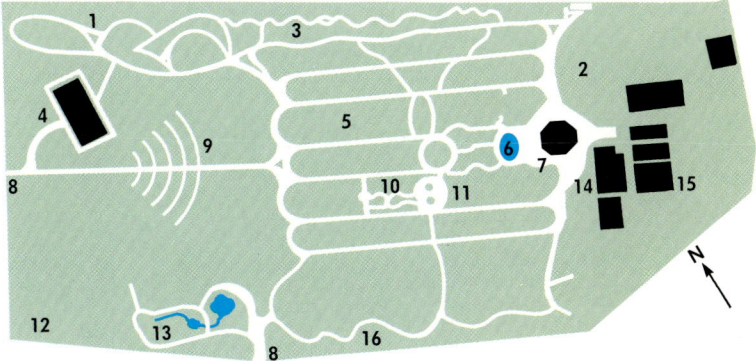

KEY TO MAP
1 Australian Plants
2 Nursery
3 Fuchsia Garden
4 Conservatory
5 Bedding Plants
6 Lily Pond
7 Geranium Conservatory
8 Entrance
9 Rose Garden
10 Bulbs
11 Herbs
12 Conifers
13 Fern Garden
14 Glass House
15 Shade House
16 Rhododendrons, Azaleas and Camellias

with the bunya bunyas. Camellias in Victoria were originally regarded as conservatory plants and grown indoors – in this case in the fernery – as they were in Europe, so the need for an outdoor camellia collection did not exist until the mid-twentieth century. The excessively alkaline nature of the soil was changed by means of loads of new soil and manure to provide the acidity necessary for camellias and azaleas.

The soil in the Geelong Gardens is a problem even today, however, as it is so alkaline that many of the plants show yellowing of the leaves despite the use of fertilisers.

Further along from the camellias and azaleas, a new outdoor fernery section has been recently built round a pond; it already features 150 species of ferns, with twenty-five species of tree ferns forming a canopy. In contrast with the early planting of the Gardens, which was arranged for landscape effect predominantly, the twentieth-century planting philosophy for the Geelong Botanic Gardens is to group plants according to their families. For many years the shrubbery along the eastern boundary in front of the avenue of *Sequoiadendron giganteum* held a fine display of fuchsias. This was recently decimated in a series of droughts, and the fuchsias are being replanted by the Friends of the Botanic Gardens. It is hoped that eventually many fuchsia species and up to eight hundred cultivars will be exhibited. There is also here, in a conservatory near the centre of the Gardens, a collection of tuberous begonias nearly up to the standard of the collection at the Ballarat Botanic Gardens. But it is for the collections of pelargoniums that these Gardens are becoming renowned. Displayed in the Florence E. Clarke House, built in 1972 at a cost of $20 000, some eight hundred varieties are grown. Potted and arranged on tiered shelves, and hanging in baskets from the roof, they cover the conservatory from ceiling to floor. Four different displays are exhibited annually of zonals (the geraniums), regals (the pelargoniums with coloured leaves), and scented species and cultivars. There is a permanent display of species pelargoniums.

Geelong also grows good roses. A new rose garden has been built near the entrance to the gardens in a fan shape, and it features modern hybrid tea roses. Set in wide lawns, the rose garden was made possible by the addition of 1.4 hectares of land brought into the Gardens complex from Eastern Park in the 1960s, nearly doubling the size of Daniel Bunce's original 5¼ (2.1 ha) acres of botanical nursery to 9 acres (3.6 ha). A large section of native plants is grown on the eastern side of the additional land and in the centre a new conservatory, the A. L. Walter Memorial Conservatory, has been built. It commemorates the work of Arthur Lewis Walter, Geelong's town clerk from 1905 to 1939.

The original entrance to the Gardens is clearly designated by the planting – probably in Daniel Bunce's time – of a pair of Irish yews, *Taxus baccata*. The fountains and statuary that punctuate the main axis of the Gardens were formerly placed in the city of Geelong. They were relocated in the Gardens in 1912, and though they are impressive and large-scaled, it is the trees within the original botanic gardens that today take the eye. Near John Raddenberry's summerhouse you can see perhaps the most famous of them, the *Ginkgo biloba*, believed to be the oldest and largest specimen in Victoria, if not Australia. With an extraordinary trunk featuring protruberances of vestigial roots, it turns a brilliant yellow in autumn and is a foil for the *Podocarpus andinus*, supposedly the largest specimen in Victoria, and the *Brachychiton discolor* planted near it.

There are some curious oaks in the Gardens, including a fine cork oak, *Quercus suber*, with a well-developed corky bark, and the valonia oak, *Q. macrolepis* var. *vallerea*, from the Middle East and the Mediterranean, which has acorns 5 centimetres wide containing tannin. In the late 1860s England imported 160 000 cwt of 'valonia' from oaks in the Middle East for tanning purposes, and in 1861 a Geelong traveller in Italy, Charles Ibbotson, sent Daniel Bunce a box of valonia acorns in the hope that Bunce would be able to acclimatise them and begin an industry in Geelong. The trees grew to 16 feet (4.9 m) in fourteen years, but their acorn crop was disappointing. The single valonia oak in the Gardens may have come from the original acorns and can be seen near the Pelargonium House. Near the cactus garden is one of the more spectacular trees of the Gardens, the Chilean soap bark tree, *Quillaja saponaria*. There is a Spanish juniper, *Juniperus*

thurifera, the only known specimen in Victoria; a fine *Dracaena draco*, at the entrance to the gardens; an avocado, *Persea americana*, which surprisingly fruits in the Geelong climate; and a Parry's nut pine, *Pinus cembroides* 'Parryana', the only one known to be growing in Victoria. There is, near the entrance, a closely planted group of native trees: a spotted gum, *Eucalyptus maculata*, of great height and distinction; a Queensland kauri, *Agathis robusta*; and a hybrid kurrajong, *Brachychiton acerifolius* × *B. populneus*. This is perhaps the most beautiful of all the groups of trees in the Gardens.

The climatic conditions of Geelong, like those of Melbourne 72 kilometres away, are among the world's best for horticulture, with an average January temperature of 24.5 degrees C and a minimum of 13.3 degrees, with a July maximum of 13.5 degrees and minimum of 5.4 degrees. There is an average rainfall of 538 mm. Low rainfall and hot winds made planting difficult for the Gardens originally, but today adequate water supplies make the most of the climatic advantages. Exotics from colder climates grow well, as do all warm-climate species except true tropical plants. Oddly, of the araucarias, only the bunyas prosper.

Native trees have always thrived in the Geelong area, and the earliest reports indicate grassland with patches of thick bush comprising *Eucalyptus*, *Allocasuarina verticillata* and *Banksia integrifolia*. The grassy hill that became the Botanic Gardens was completely bare of trees as it had been immediately cleared by the early settlers needing firewood for their lime kilns on the coast.

On 25 January 1848, after a decade of European settlement, the local newspaper, the *Geelong Advertiser*, suggested that the town was far enough advanced 'for the formation of public walks and horticultural gardens'. A few days later Charles Joseph La Trobe, the superintendent of the Port Phillip District, visited Geelong, and it is reported that he 'selected fifty-five acres, which he intends recommending to the Government as a reserve to be used for public walks and botanical gardens'.

This was the Western Gully, today the central part of the city of Geelong and encompassing the land between Gheringhap Street and La Trobe Terrace; it includes Johnstone Park and extends towards the bay. Over a year later nothing had been done about making a botanic garden, although La Trobe had approved a sum of money for the fencing and laying out of one. On 27 August 1850 the 'select committee for a botanic gardens' met to suggest that the site chosen by La Trobe was

owing to the direction in which the commercial section of the town was expected to extend, . . . utterly incompatible for any purpose that would interfere with the extension of streets and buildings in that Quarter . . . But in addition . . . soil, aspect and character of the surface [were] unfit for any purpose for which a Botanic Garden is wanted.

They recommended instead a site further from the town on the eastern side near the sea, with suitable undulating land, soil, position and good shelter. This, the present site of the Gardens and Eastern

Left: **The *Jubaea chilensis* stands on the site of the large fernery, which was demolished in 1959.**

Above: **Fountain surrounded by the cactus garden.**

Park, had been reserved by Charles La Trobe for out-of-town parkland.

An accomplished amateur botanist, Charles La Trobe had chosen, in 1846, the site for the Melbourne Botanic Gardens, but he did not oppose the Board of Trustees in their choice of Lime Burners Point, which was officially reserved in 1851. By 1854 the Victorian Government had voted £1500 to begin work on the new Botanic Gardens. The carriageways of today's Eastern Park were made particularly wide. The area was fenced in 1856, and a year later Daniel Bunce was appointed curator of the Botanic Gardens at a salary of £300 a year.

Another significant botanist, Eugene Fitzalan, who had migrated to Geelong from Ireland in 1849, had attempted to found Geelong's Botanic Gardens as early as 1855 on a property of 30 acres (12.1 ha) on the banks of the Barwon River. He had laid them out as a pleasure garden, intending 'to supply the want of a botanic Garden to Geelong'. Six months later he became insolvent and his stock of fruit trees, flowers, bulbs and herbs were sold up. He later settled in Brisbane, then Bowen, and in the early 1880s established botanic gardens in Cairns, which became today's Flecker Botanic Gardens.

Daniel Bunce had also known failures before his ultimate success in establishing Geelong's Botanic Gardens. He had arrived in Hobart from England in 1832 as an accomplished botanist and horticulturist. In Hobart he botanised round Mount Wellington and became a consultant botanist to the colonial government. He worked in a private nursery, setting out exotic and native plants in systematic Linnaean and de Jussieu beds. He eventually bought the nursery, called it the Denmark Hill Nursery, and published a monthly garden magazine known as *Bunce's Manual of Practical Gardening*.

But by 1839 the nursery had failed and was sold, and Bunce sailed for Melbourne. After exploring the Dandenongs botanically in the company of three Aborigines and their wives, in February 1841 he approached the local authorities with a suggestion that he start a botanic garden near the police paddock. This was five years before Melbourne's Botanic Gardens were eventually started. Bunce would supply all the plants and have the authority to sell crops and nursery stock from them in exchange for opening them to the public and maintaining them to the standard of those in Sydney. He wanted a twenty-five year lease. The offer was refused, but it is thought that his keenness may have prompted the eventual choice of a site for the Melbourne Botanic Gardens. In 1845 Bunce joined Ludwig Leichhardt's expedition into the Sturt Desert, and during the six months away he collected a thousand native species of plants, three of each. Leichhardt did not include Bunce on his second expedition, which left in March 1848 and became subsequently lost.

Daniel Bunce applied in 1846 for the position of superintendent of the Melbourne Botanic Gardens after the death of John Arthur, but John Dallachy was chosen. Bunce then became a freelance gardener, landscape designer, horticultural writer and explorer, botanising along the Murray River. In 1852 he was appointed manager of a mine in Bendigo and in 1857 accepted the position of curator at the Geelong Botanic Gardens. He began with such zest that within six months the *Geelong Advertiser* published the comment, 'The Botanic Gardens are beginning to assume an appearance to justify their name'.

Soon eight thousand stock plants were rooted in the nursery, and belts of young trees, four deep, were planted in the parkland. Despite a scorching January in 1860, when overnight, the foliage of newly planted trees turned to dry powder, and many smaller plants were lost, by 1861 the Melbourne correspondent of the *Chronicle* wrote of the Gardens: 'They really are beautiful and ours are not a patch upon them. You have more floral beauties in the one corner of your conservatory than we have in the whole of ours.'

The plants in the Geelong Gardens were labelled, and in 1860 Bunce published a second plant catalogue listing four thousand species. He exchanged plants with botanic gardens and nurseries throughout the world and received seeds from Germany and India. He sent seeds to the Melbourne Botanic Gardens, and four hundred poplar trees, as

well as receiving stock from their director. The Sydney Botanic Gardens also sent rare and new flowers to Geelong and, as Daniel Bunce had had a conservatory and greenhouse built and seven gardeners employed by 1859, he was ready to receive them. Bunce was able to say, showing visitors round the Gardens, 'These came from seeds I collected with Leichhardt . . .'

Bands played in the Gardens, and the *Geelong Advertiser* of 8 December 1866 commented on 'the way in which art has been made a willing slave to nature by the curator of the Botanic Garden'. So great was Bunce's acclaim that in 1868 he was asked to lay out the Botanic Gardens in Colac, 38 acres (15.4 ha) on the shores of Lake Colac. This was a site with fine native vegetation consisting of phyllodinous acacias and prostanthera, both of which he used in the design. He applied for the position of superintendent of the Adelaide Botanic Garden when its founding superintendent George Francis died in 1865, but Richard Schomburgk was appointed instead. Commented the newspaper, the Geelong *Advertiser*, 'Local influence has prevailed over distant merit'.

Throughout his lifetime Bunce's botanical qualifications were open to suspicion, for much of his writing was plagiaristic, but there is no doubt about his horticultural skill. He tried economic crops in the Gardens, including arrowroot and cotton seed. Tobacco was also tried, and continued to be grown in the Gardens until 1932.

Daniel Bunce planted many street trees, in the town and in the gardens. They were mainly blue gums, which he defined as 'The tree of trees'. He planted trees throughout the Gardens and the town, and was influential in the success of the Agricultural and Horticultural Show exhibits of flowers – particularly in 1867 when his exhibits included forty-five flower varieties sent by Veitch's Nursery from Chelsea, London. Said the *Geelong Advertiser*: 'To Mr Bunce alone are we indebted for making our autumn show a horticultural one in fact as well as in name.' Royalty visited the Gardens, and the Duke of Edinburgh planted the *Sequoiadendron giganteum* near the entrance. Birds were also acclimatised there, including sparrows.

The government supported the Gardens with grants of £1500 a year, then £2000 annually. Plants were arranged systematically according to the natural order of botanical relationships. Bunce distributed plants raised in the Gardens to public institutions throughout Victoria: in 1859, for instance, 29 576 trees, 1035 trees and shrubs, 1505 packets of seeds and 2120 cuttings were dispatched.

But by 1866 the funding had been reduced by the government; £100 was taken from Bunce's salary (originally £300), and the annual grant was reduced from £2000 to £1000. The classification of plants had to be abandoned. But the people of Geelong were pleased with the Gardens. In nine years, as the papers reported, the oak trees had borne acorns and the chestnuts had blossomed. In 1868 a great avenue of trees, which is there today, sweeping through the curves of Eastern Park, was planted the length of the Gardens. The back rows consisted of Lambertiana cypresses, and the front rows dwarf pines and evergreens.

In 1872 Daniel Bunce died, aged 59, and was buried in Eastern Cemetery, which he had earlier planted with Norfolk Island pines and cypresses. His successor, John Raddenberry, was a gardener whose parents and grandparents had been gardeners in England. He had a less scientific background than other applicants, having worked in Ballarat as a seedsman and gardener after migrating in 1853 from England.

Today not much is left of the work Raddenberry did during his curatorship between 1872 and 1899. He eliminated many of the paths and walks in the outer part of the Gardens – the Eastern Park – and he thinned out Bunce's thick plantings of blue gums. He catered for the taste of the day with an aviary, a fish pond, and in 1876 designed a new conservatory with bluestone foundations. Tuberous begon-

Opposite: Lilac is a spring favourite in the Gardens with sixteen varieties flourishing.

Right: Shrubbery path edged with 50-year-old box hedges.

Below: The camellia and azalea garden in the shade of an avenue of *Araucaria bidwillii*, bunya pines.

ias were among the plants grown there. Water was laid on to the Gardens, and they provided facilities for fish acclimatisation when the Geelong and Western District Fish Acclimatisation Society was formed. Plants were donated by both private collectors and botanic institutions, including the Botanic Gardens of Sydney, and John Raddenberry was distinguished horticulturally by his raising of a rare hakea, *Hakea grammatophylla*, from plants and seeds given him in Adelaide. (This plant is now called *H. multilineata* var. *grammatophylla*.)

But his major work was the design of the immense fernery made from slatted timber and begun in 1885. It had a central section with an arched front, and two wings in Gothic style. Halfway down the middle section was an octagonal spire that towered 60 feet (18.3 m). Completed in 1887, the fernery cost £600. Its entrance was planted with tree ferns and fuchsias hung with creepers that included cloth-of-gold roses. Down the length of the fernery were footpaths overhung with bamboo baskets of moss and ferns, and at regular intervals a thousand specimens of staghorn ferns. The sides of the fernery were planted with geraniums and at each supporting post there was a fuchsia. Beds of planting within the fernery were edged with a double row of rocks filled with small ferns.

Large beds of varying shapes contained a fairly formal arrangement of plants. Palms rather than ferns predominated, and one newspaper article suggested that 'Palmery might be a more apt name for it than Fernery'. Some beds had fishponds at either end, or miniature ponds with aquatic plants surrounded by rockeries. One had a pair of alabaster pillars of 'handsome proportions'.

Besides the *Jubaea chilensis*, which had been planted earlier by Bunce, one of the more notable plants was a *Gunnera chilensis*, like a giant rhubarb. There were camellias, azaleas and rhododendrons and other 'choice flowering plants'. There were also maple trees, sixteen varieties of holly and a flame tree, *Brachychiton acerifolius*. The building was watered by fountain jets, and the footpaths, 4 feet to 6 feet (1.2 to 1.8 m) wide, were asphalted. A guidebook published at the time suggested that the whole fernery was 'unrivalled in the southern clime'.

The fernery became an important town building, with, in December 1922, a mayoral reception of 1500 taking place in it.

Floral displays within the Gardens were elaborate and costly. In 1894 there was an exhibition of eighty varieties of chrysanthemums, which was then replaced by a display of three hundred rose plants planned to last only one season. The displays of cinerarias, geraniums and cyclamen were 'well worth a visit'. There was criticism in the municipal council, which administered the Gardens, that Raddenberry was losing his memory, and two years later he resigned.

His successor was a horticulturist called James Philip Day, who was appointed at a salary of £135 a year. During his curatorship the famous maze of

Top: One wing of the giant fernery made from slatted Australian hardwoods which dates from 1885-1887.

Above: Interior of the fernery showing staghorns and palms.

Right: The original summerhouse designed by John Raddenberry, the superintendent, in the late 1880s still stands.

the Gardens was began, in 1896. Located just inside the entrance, it was laid out in such a complex pattern that visitors had to walk a quarter of a mile to get to the centre. But the boxthorn plants lacked water and refused to grow, so that six years later its height was only 18 inches (46 cm). It became an embarrassment. 'Every town has to have a joke', commented one newspaper. Geelong's was its maze. It was eventually removed in September 1930. (By contrast, the maze built in the Ballarat Botanical Gardens was so successful that when it had to be removed to make room for the conservatory, another was designed.)

By the first decade of the twentieth century there were qualms about the correctness of the site for the Botanic Gardens, as their position on the outskirts of town, although on a tram route, was not easily accessible to the general public. A 1911 photograph in *News of the Week*, of the entrance to the Gardens, bore the caption, 'Scene in the Eastern Park, where visitors seldom go'. Advisers were called in by the Council. Christopher Mudd, who had been at one time the official botanist to King

Left: **Autumn foliage of the *Ginkgo biloba*, believed to be the oldest specimen in Australia.**

Right: **Tall palms originally planted in the fernery.**

Edward VII, and was employed by the Ballarat Water Commission as a forester and ranger, declared in 1910:

Your flower display is at the back door. Bring it to the front. The Railway Terrace and Johnstone Park, with rich green sward and some fine tree avenues, should be merged into one. Here you would have your conservatories, ferneries and plant houses. Not where very few people see them. To win the race, you must make Geelong attractive at its front door.

Christopher Mudd's opinion was backed by William Guilfoyle, who had just retired from his directorship of the Melbourne Botanic Gardens and, on his way back from the remodelling of Bunce's Botanic Gardens at Colac in 1910, stopped in Geelong and expressed surprise at finding such an ideal spot as Johnstone Park (6 acres [2.4 ha] of land) in the very heart of Geelong.

The council, however, was worried about the cost of engaging Guilfoyle as a consultant, and de-

cided to ask the advice of John Lingham, the curator of the admirable Ballarat Botanical Gardens. When Lingham recommended more flower beds for Johnstone Park the council agreed, and the Botanic Gardens of Daniel Bunce's remained 'at the back door', on the eastern side of the town.

Today the Geelong Botanic Gardens are regaining popularity, and attract an estimated 200 000 visitors annually.

The present director, Ian Rogers, sees the role of botanic gardens throughout Australia as being a source of basic collections of planting suitable for specific locations and climates – Geelong, for instance, has the finest pelargonium collection in the Commonwealth. Although he will strengthen the original Bunce design and the informal shrubbery collections and trees within the tightly packed original part of the Gardens, he aims to add more land from Eastern Park, behind the bunya bunya pines, to double the present size. The conifer section is to be extended into the new land with planting in 1986 of thirty-six trees including more *Agathis* species and araucarias, and there will be a section of deciduous temperate-zone trees to provide autumn colouring, which was lacking in the original Gardens planting. Some of the existing *Pinus radiata* will be removed, and the Geelong City Council, which has managed the Botanic Gardens since 1974, has organised expansion on a twenty-year plan to bring it well into the twenty-first century. Yet the intimate character of the gardens, with the beauty of the original shrubberies and specimen trees, and the professionalism of the design, will be retained, maintaining the distinction they have had since the first decade of Daniel Bunce's curatorship.

BALLARAT BOTANICAL GARDENS

By the 1870s and 1880s, Ballarat in Victoria was perhaps the largest gold city the world had ever known before the rise of Johannesburg. The Ballarat Botanical Gardens were established in 1858, several years after the first strike of gold, and they mirror the dignity, elegance and prosperity of the city in the late nineteenth century.

They comprise 6.7 hectares and are part of a 32 hectare botanical parkland on the outskirts of West Ballarat adjoining Lake Wendouree, a 242.8 hectare body of water stretching to the horizon. With verges covering nearly 81 hectares that are planted with willows, poplars and conifers, the lake is a particularly Oriental setting for which the Botanical Gardens, with their statuary, big trees, green lawns and brilliant bedding are a bright and mannered centrepiece.

The Gardens land is flat and the trees are nearly all exotic, with an equal distribution of deciduous temperate species and evergreen conifers planted either as single specimens, in avenues or in formal pairs. Most of them are over a century old, with forty-six listed in the National Trust's Register of Significant Trees. These include an avenue (counted as one item) of twenty-eight *Sequoiadendron giganteum* stretching 1515 metres through the centre of the Gardens – in 1874 when it was planted it was the longest avenue in Victoria; two specimens of the weeping pagoda tree, *Sophora japonica* 'Pendula', which is common as a specimen tree in China and Japan but extremely rare in Australia; a pair of the horizontal weeping Scotch elms, *Ulmus glabra* 'Horizontalis'; the only known specimen in Victoria of the Exeter elm, *U. glabra* 'Exoniensis'; and two rare Wredt's elm, *U. × hollandica* 'Wredei'.

As you approach the Gardens through the botanical parkland down beside the lake, you pass through a long avenue of large conifers – *Cedrus deodara* and *C. libani*, *Pinus halepensis* and *P. radiata* – interspersed with light deciduous trees (mostly *Populus alba*) in an impressive foretaste of the grandeur of the trees to come. Many of the trees were sent from the Melbourne Botanic Gardens by the director, Ferdinand Mueller, a close friend of the first gardener at the Ballarat Botanical Gardens, George Longley. Others came from Daniel Bunce, the curator of the Geelong Botanic Gardens.

At the entrance, opposite the lake, you notice the statuary. Unlike the reproductions of classical art that make up most late nineteenth-century sculpture in large botanic gardens, it consists of original works of art in Carrara marble donated by prominent Ballarat citizens. Thomas Stoddart, a miner and stockbroker, was so impressed with Italian sculpture during a visit to Italy in 1882 that he bought a collection of a dozen original statues in the classical manner by the leading Italian sculptors of the day to place in his 'beloved' Ballarat Botanical Gardens. Today, randomly sited among the trees, they are among the signatures of the Gardens. Impressed by Stoddart's gift, a fellow miner, James Russell Thomson, bequeathed the money to purchase classical marble works of heroic size from Australian sculptors. These were to be housed in the elegant statuary pavilion in the Gardens. Richard Armstrong Crouch, a solicitor and soldier, politician, art patron and historian, also left a bequest for statuary, stipulating that the works be busts of each Australian prime minister. The first six busts were unveiled in 1940. The collection has been continued, and is now the responsibility of the Ballarat City Council. An MLC, the Hon. D. Ham, in 1893 contributed the pair of marble lions that still stand inside the fine entrance gates donated by the Hon. E. Morey, MLC.

The architecture in the Gardens is High Victorian and of the best quality. The statuary pavilion, unique in Victoria and probably Australia, was designed by T. E. Molloy in 1887 and has an octagonal plan, a corrugated iron roof and cresting eaves. It is topped by a small octagonal lantern light, with its own cresting eaves, and by a miniature dome with a decorative finial. All the faces of the building have large panes of glass flanked with narrow strips of glazing in the Regency style. Inside the pavilion one piece of sculpture is displayed in each of four corners, and in the building's centre, protected from the public by brass rails, is the massive group depicting the flight from Pompeii. There are two picnic

---

Left: **The statuary pavilion in the Ballarat Botanic Gardens is one of the many fine examples of park architecture in the Gardens.**

pavilions that demonstrate the best in Victorian municipal design. The Almieda Pavilion, on the Lake Wendouree side of Wendouree Parade and opposite the Gardens, was originally an amusement pavilion with amusement machines at its centre. The other pavilion within the Gardens was built in 1904 and has a concave roof of corrugated iron; it is almost Japanese-looking and is held in place by ten iron pillars with latticework sides to half the wall height and the rest is open to the air.

The Gardens have a large timber-slatted fernery planted with tree ferns and smaller ground-level species in lozenge-shaped beds. The third fernery to stand on the site, it is much simpler than the original fernery and camellia house built in the late 1870s which, according to the mayoral report of 1884 was 'admittedly unique' and by that time represented 'a splendid horticultural interior: A garden flourishing at all times irrespective of season and equally defiant of hot winds and winter frosts.'

Ballarat's climate is suited to the temperate flora grown in the Gardens. While temperatures may reach the high thirties in summer, the nights are invariably cool and pleasant. Winter temperatures vary from minus 4 or 5 degrees C overnight to 10–12 degrees C through the day; there is an annual rainfall of 675 mm.

Tuberous begonias have been a speciality of the Gardens since the 1890s. In the autumn there is an annual Begonia Festival featuring about ninety species of tuberous begonias in the cold house conservatory. Visitors also come to the Gardens to look at the display of annual seedlings, bedded out geometrically in the lawns. In two annual displays, a total of twenty thousand seedlings raised in the Garden's own nursery are used.

There are about three thousand species in the Gardens, ranging from groundcovers to the giant sequoias; many are labelled. Its role though is primarily that of a pleasure garden rather than a scientific one. A mayoral report of 1888, noting that a complete system of labelling to facilitate recognition of trees and shrubs for prominent specimens had begun, added: 'with all due respect for scientific study, the Mayor is of the opinion that the general public will always care for the gardens more as a resort for recreation and pleasure.'

The attraction of the Gardens was always linked with that of Lake Wendouree. Anthony Trollope, visiting Ballarat in 1871 to gather material for his book *Australia and New Zealand*, was more impressed with Ballarat than with 'any other city in Australia'. He wrote:

It is so well-built, well-ordered, endowed with . . . schools, hospitals, hotels and every municipal luxury that can be named including a public garden full of shrubs and flowers and a Lake of its own – Lake Wendouree – with a steamer, row boats and regattas.

Abundant natural water had been one of the attractions of Ballarat when it was the main meeting-place for Aboriginal tribes of the Central Highlands area of Victoria. In the 1840s the fertile wooded land and the water attracted pioneer farmers including Archibald and William Yuille, who settled in the area of swamp near the edge of Lake Wendouree that became Yuilles Swamp.

After the gold strike of 1851, the influx of miners from all over the world, and the proclamation of the town in 1855, the swamp supplied the town's drinking water. In 1857 the Police Paddock of 80 acres (32.4 ha) on the west side of the swamp was reserved for the Ballarat Botanical Gardens, and a year later a road was built around the swamp. Developed simultaneously, the Gardens and the lake became the major pleasure grounds of Ballarat. The lake was cleared of weeds by prisoners, its margins were firmed by the dumping of loads of reef quartz from mining, and exotic trees were planted 'by the half hundreds'. English grasses for mown lawns were sown on 4 acres (1.6 ha) along the lake verge facing the Gardens, and walkways, bridges and boating piers were built. In 1861 swans were donated – but a donation of native bears was refused as the 'time had not yet arrived for a menagerie'. In 1862 fish for

KEY TO MAP

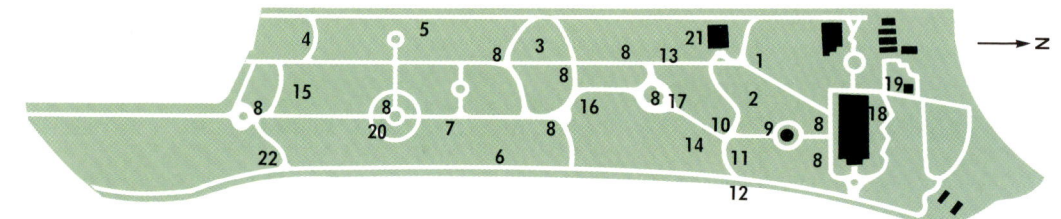

1 T.B. Toop Corner
2 Thomas Beaumont Lawn
3 Azalea Garden
4 Laburnum Arbour
5 Rock and Alpine Garden
6 Floral Clock
7 Avenue of 28 Big Trees
  (*Sequoiadendron giganteum*)
8 Thomas Stoddart Statues
9 The Statuary Pavilion
10 Sir William Wallace Statue
11 Marble Lions
12 The Gates
13 Prime Ministers Avenue

14 Reflection Bowl
15 Betty Johnson Rose Garden
16 Rose Gardens in honour of
  Lady Baden Powell
17 Wishing Well
18 The Fernery

19 Adam Lindsay Gordon's
  Cottage
20 Tilley Thompson Garden
21 Conservatories (Begonia
  Houses)
22 McDonald Pergolas

the lake, particularly trout, were reared by the Ballarat Fish Acclimatisation Society in a hatchery in the Botanical Gardens. A 95 foot (29 m) ornamental house, two aviaries and a pond for wildfowl were built in the 1870s on the lake verge. Today, vestiges of the full-bodied pleasure ground developed by the 1870s opposite the Gardens remain in the bridges, trees, lawns and walkways of 'Fairyland', and in the Lake Lodge, the Gardens' restaurant, which was designed as the result of a competition and built in 1900.

A competition for the best design for botanical gardens at Ballarat had been held in November 1858, with a prize of £10 for the winners, Messrs Wright and Armstrong, of whom nothing is now known. A month later the Ballarat Municipal Council advertised in the newspapers for a gardener, to be paid £3 a week, to begin making the Gardens. From fifteen applicants George Longley, who had been an apprentice gardener at Lowther Castle in Westmoreland, England, was chosen. He had been trained by the resident landscape gardener at Lowther Castle, George Cruickshank, who was a well-known landscape designer and gardener of the day, and who had also done the illustrations for *Pickwick Papers* as he was a friend of the author, Charles Dickens.

Pitching a tent under a gum tree in the new Botanical Gardens, approximately where the residence now stands, George Longley began to clear the swampy, scrubby ground. He commenced a term as gardener, for many years the only gardener, at the Botanical Gardens that lasted till his death in 1899. He had come to Ballarat as a miner in the first gold rush, but having been unsuccessful went to Melbourne where he read the advertisement for the Ballarat Botanical Gardens. He is reputed to have chosen the site for the Gardens, and to have been a close friend of Ferdinand Mueller, who was a frequent guest at the Longley family home and was vitally interested in the progress of the Gardens.

Early in 1858 the first grant of £500 from the Victorian Government had been approved, and a committee of management was appointed to co-operate with the municipal council in the laying out of the Gardens. Contract labour was hired to clear the site of trees and stumps, and loads of soil were taken out of the swamp to build up the land. By 1860 plants and seeds were being received from the Melbourne and Geelong Botanic Gardens and from private citizens. One consignment from Ferdinand Mueller is recorded as having to be returned as it contained not only species which the Gardens already had, but ones of inferior quality. There were, though, consignments of excellent stock. Seven and a half acres (3 ha) of the Gardens were fenced and a quarter of an acre (100 sq.m) was reserved as a propagating area. But 1860 was a bad summer for the Gardens, when a plague of grasshoppers destroyed most of the planting.

A maze of *Acacia paradoxa* was planted in 1862 on the site of the present cold house conservatory. Planning for the planting of the *Sequoiadendron*

giganteum avenue began in 1862, with tree guards being erected and the soil dug, but final planting did not take place until 1874. The avenue, which stretched into the parkland at both ends of the Gardens area, is today unchanged except that it then had a footpath 12 feet (3.7 m) wide and verges of English grasses half a chain (10 m) wide on either side of the trees.

By then two windbreaks on the south and north sides of the Gardens had been established, a rotunda had been built, and drainage had begun. Labelling of prominent trees had started in 1864, and 7 acres (2.8 ha) of the Gardens had been sown with English grasses, 2 acres (0.8 ha) with forest trees and shrubs.

In the late 1880s the larger buildings for the Gardens were begun, with a big conservatory planned for the old maze site and designs from England invited. A new maze was opened in 1888. The statuary pavilion was built in 1887, and the statues of Thomas Stoddart's choice were in place in 1884. A rockery, pond and aquarium were built at the western end of the fernery. The straight avenues of trees edged with bedding plants, thought to be a bit monotonous, had been relieved by increased lawn-planting and a less formal arrangement of new trees.

Unwanted trees were by then spoiling potential vistas and were cleared away so that, as the council minutes recorded, 'Grass could be sown to provide in every direction, green vistas of well-cared-for lawn, broken only by the very best specimens of trees and shrubs'.

Left: A newly installed 19th century fountain is in keeping with the period of the Gardens.

Above: The Avenue of Australian Prime Ministers' busts was begun in 1940.

Right: Brilliant beds of annuals are a spring attraction of the Gardens.

Below: The large conservatory seen from the Avenue of Prime Ministers.

There were also flourishing zoological gardens in the north-western corner of the Gardens. By the 1870s, owing to donations of deer and native birds and animals, George Longley had been forced to open a zoo, and some of the aviaries had been transferred across the road from the lakeside to the Gardens. By the 1950s the zoo and most of its concrete structures quietly disappeared from the Gardens.

Today the design is much the same as it was when it was laid down in the 1858 plan. The specimen trees have not changed much. To the rear of the western border of the Gardens a row of variegated silver and gold *Acer negundo* trees forms a new screen from the road and a background for a shrubbery bed that runs the length of the Gardens. Instead of the palm lilies, *Cordyline australis*, of the 1870s and 1880s in the area where the busts of the prime ministers stand, an avenue of flowering chestnut, *Aesculus hippocastanum*, frames the sculptures. The Wellingtonia avenue (that is, the avenue of *Sequoiadendron giganteum*) is still the dominant feature of the Gardens and a dark foil for the deciduous light beauty of ashes, elms, beeches, poplars, and a pair of *Taxodium distichum* that frames the statuary pavilion.

Above: **The tuberous begonias of the Ballarat Botanic Gardens are famous and can be seen in the autumn.**

Right: **Cyclamen fill the cold house conservatory in the spring.**

Centre right: **Alpine plants grow in the new Bicentennial rockery in the Gardens.**

Far right: Deutzia nikko in the rock garden of the Ballarat Botanic Gardens.

The curators of the Gardens since George Longley died have all implemented the original policy of maintaining them as an ordered, beautiful pleasure garden. George Longley's immediate successor was John Lingham, who had been in charge of the south section of the Gardens during the last years of Longley's curatorship. Thomas Rooney was in charge of the north section. He in turn succeeded John Lingham, and in 1914 Thomas Toop, who had already been a gardener employed at the Ballarat Botanical Gardens since 1890, was appointed curator. The son and grandson of English gardeners, he remained at the Gardens as curator until he retired in 1945. He had trained his successor, Thomas Beaumont, from a 'lad gardener' of thirteen years and nine months. Beaumont started work in 1924 and retired in 1978. Thomas Beaumont succeeded Thomas Toop as superintendent in 1947, having made a world-wide reputation for his culture of tuberous begonias. The two main lawns of the Gardens near the fernery end are named the T. B. Toop Corner and the Thomas Beaumont Lawn, respectively.

To keep the Gardens in top condition despite frequent bad droughts, scorching summers and cold winters requires horticultural expertise. The current Director of Parks and Gardens for the City of Ballarat, who administers the Botanical Gardens, is Robert Whitehead. He has instituted the first major botanical addition since the days of George Longley and Ferdinand Mueller with the building of a rock and alpine garden. This was made possible by funds from the Victorian Government, given to commemorate the 150th birthday of Victoria, and by money from the Cuthbert Bequest, and it displays some seven hundred named species of small alpine and rock plants. The framework for the garden is a variety of conifers and a *Corylus avellana* 'Contorta', which is an unusual and interesting small tree. These and a number of other plants in the new

garden were provided by local residents and the Victorian Alpine Plant Society.

The rock garden was endowed with good local rocks, some of which are now covered with *Scleranthus biflorus* and *Acaena microphylla*, and pockets of soil are planted with alpine plants including *Gentiana acaulis*. There are also dwarf gladiolus, a dwarf *Kowhai* (*Sophora tetraptera* 'Gnome'), dwarf hebes, a dwarf broom and various irises, some miniature forms of epimedium, and *Convolvulus cneorum* and *Armeria maritima*.

A laburnum arbor has been planted, and a fountain (c. 1861) from Europe has been provided for an existing circular pool. A local specialist in producing unusual and replica steelwork has manufactured lights and light stands in keeping with the period of the Gardens.

The azalea and rhododendron section, some of which was planted in the 1890s, has been recently refurbished with new varieties, and a grove of young silver birch, *Betula pendula*, has been planted. You enter this part of the Gardens from one of the earliest-planted sections behind the fernery. Retaining a nineteenth-century romantic woodland atmosphere, it has paths lined with mossy rocks backed by large specimen trees. Here you can see the better of the two weeping pagoda trees, *Sophora japonica* 'Pendula', and close by is a remarkable poplar, *Populus deltoides*. An informal and intimate part of the Gardens, it is a contrast to the vast open spaces that characterise the main part.

The essence of the Gardens is today nevertheless still that of the late nineteenth-century pleasure parks that are found in most Australian provincial cities, but on a grand scale and of the very best quality. There are the exotic trees beloved of the nineteenth century, the elegant pleasure garden architecture, the smooth green lawns broken with statuary, and the brilliance of annual bedding plants.

Top left: An *Araucaria bidwillii*, the bunya pine, and other conifers seen from the rock garden.

Left: Large trees, straight paths, statuary and bright annuals are hallmarks of the Gardens.

Above: Fairyland, a 19th century pleasure-ground on Lake Wendouree, adjoins the Gardens.

The historian Geoffrey Blainey, in his foreword to a book on historic Ballarat, writes: 'Perhaps in no other city of Australia is the late nineteenth century so persuasive . . . Old Ballarat is a vital national asset.' The Botanical Gardens, with their pleasure garden atmosphere and 'fairyland' Lake park, are part of that asset.

6
BOTANIC GARDENS IN BRISBANE

The Brisbane Botanic Gardens, or the Brisbane City Botanic Gardens as they are now called, are just 1 kilometre from the city's GPO. Flanked by the Brisbane River and the main city streets, Alice, George and Edward, they lie under the shadow of the Parliament House of Queensland.

With about 19 hectares of sweeping lawns, perpendicular palms, elegant avenues of dark trees, specimen trees, both bizarre and beautiful, ponds and statues, and curved roads along which cyclists and pedestrians pass, they are the model of a Victorian-age pleasure park. The design is unremarkable but pleasant and the planting, though graceful, is sparse. One reason for the latter is that over the years many plants have been lost through flooding of the Brisbane River. Another reason is that some of the important collections (such as the succulent collection) of plants have been moved to a second garden, opened in 1976 at Mount Coot-tha, 5 kilometres from the city centre. The recurrent flooding of the city Gardens, and insufficient space to allow for the variety of plants that the subtropical climate of Brisbane will support, led to the establishment of Mount Coot-tha Botanic Gardens. A decade later Mount Coot-tha grows 14 000 species of exotic and native plants in a natural bushland setting of 70 hectares, and has become the official botanic garden of Brisbane.

The climate, with its mean maximum January temperature of 28.9 degrees C and minimum of 20.9 degrees, its July mean maximum of 20.4 degrees and minimum of 9.8 degrees, its high level of summer humidity and annual rainfall of 1125 mm, has the potential to support the widest range of plant species of any Australian botanical gardens.

Established with local and federal funding of about $10 million in 1970, the Mount Coot-tha garden has had more generous financial support than Brisbane's first Botanic Garden, which Harold Caulfield, a former curator–director called 'The Cinderella of Australia's Major Botanic Gardens'. An earlier curator, Philip McMahon, who was appointed in 1889, wrote in his report to the Department of Lands, which controlled the Gardens at the time: 'Sir I have the honour to present my annual report, I trust that you will excuse the brevity of it, for you in your wisdom have not allowed me sufficient money to purchase note paper . . .' The curator's cottage and the superintendent's residence were constantly being flooded by the Brisbane River, and not only were the living quarters damp, causing ill health for the staff, but there was a loss of records. The herbarium was moved from a department of government to the Gardens and back again, and the administration of the Gardens changed from government to municipal.

But the history of the Brisbane City Botanic Gardens, embodied in their trees, makes them nationally significant. Here were planted the first specimens of economic crops (for example the macadamia nut) and decorative trees (such as the jacaranda) that were to become commonplace throughout the continent and, like the gardens of Sydney and Hobart, they initially provided food for the convict settlers.

In mid-1828 Charles Fraser, the colonial botanist of New South Wales, and Allan Cunningham, the official botanist from the Royal Botanic Gardens at Kew, explored the banks of the Brisbane and Logan Rivers. On 1 July they arrived at Brisbane Town, where Charles Fraser had been directed to 'establish a public garden . . . to collect the vegetable products of the country and to make observations on their uses and importance especially of the forest trees and to report on the nature of the soil and to what extent it is fitted for agricultural purposes or grazing'. Allan Cunningham hoped to collect new specimens for the Royal Gardens at Kew and to explore the area from Brisbane Town to the Darling Downs.

On 2 July they chose the site for the 'New Garden' on a bend of the Brisbane River, where they felled a 'magnificent tree of Flindersia australis', the crows ash, laden with fruit. In the morning they explored the forest on the other side of the river and Allan Cunningham collected a new specimen – the staghorn fern, *Platycerium grande*, the seeds of which were subsequently tried at the Glasgow Botanic Garden. Further up the river from the town they found the banks clothed with eucalypts,

Left: The Walter Hill Memorial Avenue of *Ficus benjamina* marks the original boundary between the Botanic Gardens and Queen's Park.

a few straggling araucarias, *Flindersia australis,* and many other timber trees, as well as a vast variety of shrubs. Closer to the town, the botanists found that the banks became more elevated and rocky.

On 3 July Fraser noted in his journal: 'Employed this day in laying down the boundaries of the New Garden and fixing the situation of a large pond in its centre.' The area was 42 acres (17 ha). Of the original vegetation in the Gardens, three large specimens of *Eucalyptus tereticornis,* on the banks of the Brisbane River, remain. Through the Gardens ran a small muddy mangrove creek, which flowed towards 'Frog's Hollow', now Albert Street.

By 1836 the 'New Garden' was described by James Backhouse, a visiting Quaker philanthropist, as 'Twenty-two acres of Government Garden for the growth of sweet potatoes, pumpkins, cabbages and other vegetables for the prisoners'. Brisbane was then under strict military control, with no free person allowed within 50 miles of the town. But in 1842 the area that became the Botanic Gardens was given to residents by the government of New South Wales, which had jurisdiction over Queensland, for the introduction of new seeds, cuttings and crop-growing. Rice was tried, but the attempt was futile as the seeds were of processed rice from a grocery shop; the climate, however, was blamed.

On 21 February 1855 Walter Hill, a 35-year-old Scot from Dumfrieshire, who had been trained for two years at the Royal Botanic Garden, Edinburgh, and from 1843 to 1851 at Kew Gardens, was appointed by the imperial government as superintendent of the Botanic Gardens at Brisbane and given £500 to spend on plants for them. He had arrived in Sydney in 1852, made his way to the goldfields of Bendigo in Victoria and become a partner of Frederick Strange in collecting objects of natural history. On an expedition to the north coast Strange was killed by Aborigines, and Walter Hill returned by ship to Sydney.

In Brisbane he was given 9 acres (3.6 ha) of the original 42 acres for the Botanic Garden, which did not quite go down to the river. Today this, the earliest-planted area of the Gardens, is known as Residence Hill, and many palms from Hill's once-famous collection of about a hundred species are growing there. In a thick rainforest section, with palms, staghorns and birdnest ferns, grow important trees: the large West Indian *Swietenia mahagoni,* the first specimen of the mahogany timber tree grown in Australia; the first macadamia nut tree planted in Australia, collected by Walter Hill at Maryborough and recommended as a 'delightful nut tree'; the first tamarind tree grown in Australia, *Tamarindus indica,* and the first lychee, *Litchi chinensis.* The first jacaranda in Australia grew here too, but was destroyed in a storm and removed in 1979.

Along the river bank from the Edward Street gate, Walter Hill planted in 1858 an avenue of twenty-eight bunya pines, *Araucaria bidwillii,* which today is one of the loveliest and most impressive parts of the Gardens. Believed to be the first bunya

pines planted for decorative purposes in Australia, they were established as a memorial to John Carne Bidwill, who had died in 1853 in Maryborough, Queensland, after being lost in the bush for eight days on a collecting mission. He was the commissioner of crown lands at Wide Bay, Queensland, a position he had nominated for himself after being deposed as director of the Sydney Botanic Gardens by Charles Moore. John Bidwill had discovered *Araucaria bidwillii,* named after him, on a collecting expedition to the north of Brisbane, and taken seeds of it to England in 1843. A relative of the internationally sought, bizarre monkey puzzle tree, *Araucaria araucana,* the bunya pine became an instant success in the northern hemisphere.

In his garden at Maryborough he grew what is believed to have been the first mango tree raised in Australia, and most of the varieties of fruit now growing in the Maryborough area probably originated in his garden. Many rare plants from the garden, including the Chinese raisin, *Hovenia dulcis,* as well as the mango, were given to the Brisbane and Sydney Botanic Gardens. Other farmers near Brisbane, at Toowoomba and on the Darling Downs, were also trying new crops, such as cotton and exotic fruits, but with the establishment of the Botanic Gardens under Walter Hill, plant introductions became systematic. A cairn in the Botanic Gardens shows where sugar cane was first grown in Australia, and it is believed that Walter Hill and a planter from Barbados, John Buchot, made the first sugar ever granulated in Queensland. They did it in the dead of night, boiling the sugar cane juice from the stalk in a saucepan to prove that locally grown sugar cane would granulate.

By 1872, under Walter Hill's guidance, the Gardens supplied to Queensland's growers 50 200 cuttings of sugar cane (fourteen varieties); 5000 cuttings of white mulberries; 5056 coffee plants; 1020 tea plants; 1060 ginger roots; 300 'papers' (presumably cuttings) of Manila, Havana and Shiraz tobaccos; and indigofera and logwood, which were produced for dyes. Figs, arrowroot, tapioca and turmeric were also grown, and palm oil, mangoes, cloves, nutmeg, allspice and cinnamon were cultivated in insufficient quantity to meet the demand for the plants.

In 1874 Hill's report stated: 'I deem it my duty to point out the great national benefit the botanic gardens have been . . .' He listed products including sugar, rum, molasses, wines and brandy, tobacco, cigars and snuff which owed their development to the Gardens. Individuals and establishments were supplied with cuttings, roots, plants or seeds, with 680 recipients in 1874.

The Gardens themselves were set out in an unremarkable manner, as Walter Hill seems to have

Top right: A *Ficus benghalensis,* one of the twenty-six *Ficus* species seen in the Gardens.

Right: The Brisbane River, seen from the Gardens.

been a plantsman of the same type as Mueller in Melbourne, favouring trees planted in straight lines rather than for landscape effect. James Veitch of the nursery firm J. H. Veitch in England, after investigating the leading botanical gardens round the world, wrote in the *Gardener's Chronicle* of 1887: 'In the Brisbane Botanical Gardens much good work is being done, . . . but I was not over impressed with the layout.'

Marianne North, the British artist who visited Brisbane during an horticultural tour of the world in 1880, substantiated Veitch's view. Staying in August with Governor Bell and his wife at Government House, the garden of which opened onto the Botanic Gardens, she found a cool wind, hot sun and dust.

The famous Araucaria trees in the Botanical Gardens were brown and dusty and not larger than the one in the temperate house at Kew. There were few flowers. Dracaenas, strelitzias and Norfolk Island Pines give a different look to the Gardens as well as the wattle trees, yellow with thousands of fairy balls and leaves mimicking the Eucalyptus.

She painted a view of the Gardens looking down on the river and over to the hills opposite, with a hoop pine, *Araucaria cunninghamii*, in the foreground and a rolling grassy paddock in which were a couple of specimen trees. Along a pathway in the foreground were square flower beds containing what look like roses.

Walter Hill loved excessive duplication of species. Shrubberies were almost non-existent; the majority of trees were single specimens in lawns. But there was a fine collection of poinciana trees, coral trees, jacarandas, palms and *Ficus*. Even today the Gardens hold one of the largest collections of *Ficus* species in Australia – twenty-six of them. There was beauty in the lagoons of the Gardens, praised in an article in the *Gardener's Chronicle* of

February 1875. Of the pond inside the Albert Street gates, it said: 'This lagoon of 206 feet by 144 feet is one of the Gardens' greatest attractions being filled with glorious water lilies, teeming with fish of many species and live with water bird life.' The four top ponds, of which only two remain, were said to be a glorious sight when the lotuses were out. Three of the lagoons each held a different-coloured lotus – white, pink and red – and the fourth contained the native Queensland large blue *Nymphaea.*'

The garden wall, which borders Alice Street and George Street, was built in 1865 with stones from the demolished first Brisbane gaol. It remains an attractive feature of the Gardens. Specimen trees planted in Walter Hill's time and which remain spectacular today are the peepul tree, *Ficus religiosa*, from India and the dragon tree, *Dracaena draco*, the first of the species in Australia. The avenue of *Ficus benjamina*, today stately and attractive, was planted in the 1870s to mark the barrier between the Botanic Gardens and Queens Park. The latter incorporated all the land in front of Parliament House to the river and was a public park with a cricket pitch. In 1916 it was officially amalgamated with the Gardens, making a total area of 48 acres (19.4 ha). The part of Queens Park in which football was once played became the palm lawn.

Walter Hill retired in 1881, having firmly established the Botanic Gardens. He had been on plant-collecting missions throughout North Queensland seeking native plants for fodder, timber and medicine or of botanical and scientific interest. Besides growing exotic, economic and decorative crops for dispersal among the Queensland population, he collected new species for stocking other botanic gardens. He helped to establish what he called the 'branch botanic gardens' of Ipswich, Toowoomba, Maryborough and Rockhampton, run by the government for the testing of economic crops, and foresaw the Daintree River area, among others, as being ideal for the cultivation of tropical fruit. He pleaded for the establishment of botanical gardens or recreation reserves whenever a township was laid out, and suggested leaving the indigenous timber intact.

Seeds from the Brisbane Botanic Gardens were sent to Hong Kong, Calcutta, Java, Glasgow, Florence, Mauritius, Paris, Washington, Sydney and Melbourne, as he noted in his 1874 annual report. They arrived at their destinations by ship and, with no allowance made for transport costs, Hill had to ingratiate himself with the various captains to take the seed packets without charge. He supplied seeds to one captain, Commander Bedwell of the *Pearl*, for planting on offshore islands, 'to afford vegetable food for shipwrecked sailors'. Pigs and goats were also left in these places for that purpose.

After Hill's retirement, the botanical side of the

Gardens diminished. The library was moved into the office of the new colonial botanist, Frederick Manson Bailey, who was appointed in 1881 and worked also in the Museum, and the Gardens were managed by James Pink, a gardener. The Botanic Gardens by this time had been eclipsed by Bowen Park, established by the Acclimatisation Society opposite the Brisbane Public Hospital. The park had been founded in 1862 to disperse plants imported from abroad, and it was well landscaped. Today it has shrunk to 2 hectares. Though Frederick Manson Bailey produced a catalogue of plants contained in both Bowen Park and the Botanic Gardens, the botanical emphasis of the latter did not recover until 1889, with the appointment of Philip McMahon as their curator.

But from 1890 to 1897 there were floods throughout the Gardens. The curator's house and office were swept away, along with the superintendent's house in 1893. A gunboat and steamer came to rest in the Gardens, and were floated off fourteen days later by a subsequent flood. The palms on Residence Hill today have trunks at 45 degree angles, the result of their being swept over in the floods of the 1890s despite attempts, using horses and winches, to stand them upright.

Above: **Some of the many tall palms in the Gardens.**

Below: **The Avenue of *Araucaria bidwillii*, bunya pines, planted in memory of botanist John Carne Bidwill, who died in 1853 in Maryborough.**

Above: One of the ponds in the Gardens. In the 19th century these held many waterlily species.

Below: The Botanic Gardens are situated close to the centre of Brisbane city.

The present curator's house, with a custard apple tree 150 years old in the garden, was built in 1903. The Gardens came under the control of the Department of Agriculture in 1888 and by 1926 control passed to the Brisbane City Council, which still manages them today.

The curatorship had passed to the son of Frederick Manson Bailey, John Frederick Bailey, in 1905. John Frederick Bailey enhanced the family's botanical tradition by becoming government botanist for the Brisbane Gardens as well as curator of the Gardens, which again became the botanical centre of the state, incorporating the herbarium and botanical library. When John Frederick Bailey became director of the Adelaide Botanic Garden in 1917, his nephew C. T. White became Brisbane government botanist and E. W. Bick the curator.

During Bick's curatorship the Gardens, owing to lack of finance, were simply maintained as a pleasure park. In 1919 the sports fields of Queens Park were filled with spoil from the newly graded river bank and the reclaimed area was planted with exotic shade trees. Today's impressive group of royal palms, *Roystonea regia*, were planted then. There was no improvement in the Gardens botanically until the appointment in 1946 of J. F. Bailey's son, J. R.

Left: A coral tree, one of the many species of *Erythrina* to be seen in the Gardens.

Bottom left: Curving paths lead through shrubberies in the Gardens. The shrub *Eranthemum pulchellum* is in the foreground.

Top right: Eucalyptus pruinosa, painted by Ferdinand Bauer during a voyage round Australia with Matthew Flinders from 1801 to 1803, from *The Flower Paintings of Ferdinand Bauer* by W.T. Stearn.

Bottom right: Hodgsonia heteroclita, an Asian climbing plant, painted by W.H. Fitch and published in 1855 in J.D. Hooker's *Illustrations to Himalayan Plants*, one of the many rare books in the Queensland herbarium library.

Far right: The Brisbane Botanic Gardens in 1880 painted by Marianne North. In the foreground is an American *Passiflora* in flower. (*Royal Botanic Gardens, Kew*)

Bailey, as curator – the third generation of Baileys to be associated with Australian botanic gardens. (J. R. Bailey had formerly been curator of Too-woomba Botanic Gardens.)

Today's Gardens owe their form and design to the curatorship of Harold Caulfield, who was appointed in 1956. He too had botanical forebears, his great-grandfather having been a horticulturist working with Ferdinand Mueller in the Melbourne Botanic Gardens, and he worked in the Melbourne Botanic Gardens and the Adelaide Botanic Garden before coming to Brisbane. He introduced three thousand new species to the Gardens, and sweeping changes in design, with a new landscape plan prepared by Harry Oakman, the then director of parks for Brisbane.

The ornamental lake system was rejuvenated, new shrubberies were built and planted, and a native plant section with some six hundred native species was begun on the banks of the Brisbane River beneath the *Araucaria bidwillii*. Up until this time the Gardens had held no native flora except the fashionable araucarias. Harold Caulfield dis-

pensed with the zoological collection, which had begun in the late 1920s with some water birds and donations of animals, progressing to include a bear pit and many more animals in the 1930s; by the 1950s it had become a motley collection of animals.

While the plantings, including a fine collection of cacti, succulents and bromeliads, were being built up, disaster struck again in 1974. Floods washed away the shadehouse, and the finest collection of bromeliads in Australia was lost. The Gardens were coated with a chemical silt and had to be closed for ten weeks. This catastrophe added conviction to the decision that the original Botanic Gardens of Brisbane were no longer adequate as a scientific exhibition ground, and that a proposal by the Brisbane City Council for a second botanical garden to be started at Mount Coot-tha should be adopted.

Mount Coot-tha is located on the eastern slope of the Taylor Range, 5 kilometres and ten minutes drive from Brisbane. Consisting originally of open sclerophyll woodland, it is 213.4 metres above sea level. By January 1976, with the designing done by Deane Miller, the Council's landscape architect, and the field work by Harold Caulfield and Barry Dangerfield, the first curator of the Mount Coot-tha Gardens. Said Deane Miller of the design:

The form of the garden could be described as 'organic'. The building complex is strictly rectilinear in form as are the areas in its immediate vicinity, but inside the Garden proper nothing is forced. Paths follow the contours in broad simple sweeps ... An over-riding principle has been that it must look right on the ground ...

No other spot near Brisbane had such natural advantages, which the staff could embellish.

The position of the Gardens is pleasant, with a view over the city but also a sensation of being 'in the bush'. The Gardens are in two sections. The 35 hectares first designed and planted, known as Stage I, include the administration buildings and the specialist gardens. The star of these is the geodesic dome housing the aquatic plants and tropical plants, with a ramp circling upwards through the building round a central tank of water plants and fish, onto which give windows showing illuminated waterscapes. There are about two hundred species of tropical and aquatic plants, including the giant waterlily, *Victoria amazonica*, from the Amazon, *Euryale ferox*, a giant waterlily from South-East Asia, and *Nymphaea gigantea* from northern Australia, are growing in the heated dome.

The dome cost $825 000 and is made of tinted acrylic bubble panels. Alongside the geodesic dome is the arid-zone garden. This contains some of the most spectacular succulents and cacti from the world's dry regions, and the collection from the Brisbane City Botanic Gardens formed its nucleus. There are about two thousand species shown from all over the world, including the kalanchoes of South Africa and the cacti of the American deserts set in a rocky landscape.

Nearby is the Fragrance Garden, with about three hundred species of fragrant plants from all over the world, including forty species of mint. Visitors are encouraged to pick the leaves of plants as they pass by and to crush them for the scent. There is a large variety of herbs, including a horseradish tree, and lemon grass, lavenders and jasmines. This is designed as an English herbaceous border style of garden and is semi-enclosed.

Stage I of Mount Coot-tha Gardens is designed primarily to exhibit the exotic flora of the world, and is suited to that of the cool-temperate zones. Further away from the administration buildings, beneath a canopy of indigenous eucalypts, the flora of the world's cool-temperate countries is arranged according to geographical groups.

There are species from Africa, Asia and the Americas that incorporate a wide variety of deciduous plants and bulbs. There are large collections of azaleas, and 250 camellias. Magnolias, too, are splendid here, with fifteen varieties cultivated. Scattered throughout the Gardens is the palm collection comprising 450 species of palms from all over the world, based on the original fine palm collection in the city gardens, and encompassing the range from cool-temperate climate species to tropical.

Stage II of the Mount Coot-tha Gardens consists of 37 hectares of native botanic garden introduced into the indigenous sclerophyll forest. This includes twelve defined areas of rainforest, ranging from the southernmost cool-temperate rainforest of Tasmania to the low scrub rainforest of York Peninsula in the north of Australia. For some of the rainforest, misting irrigation has been introduced. The rainforest planting follows a series of waterfalls, pools and watercourses (designed in the reserve by the first curator, Barry Dangerfield), which look entirely natural but which all connect so that the waters are recirculated. There are dry areas within the reserve with the flora of Western Australia, the Kimberleys and the Mallee country represented. Already about 2500 Australian native species have been established within the natural forest and work is continuing.

Right: *Aloe arborescens* from South Africa in the arid zone section at the Mount Coot-tha Botanic Gardens.

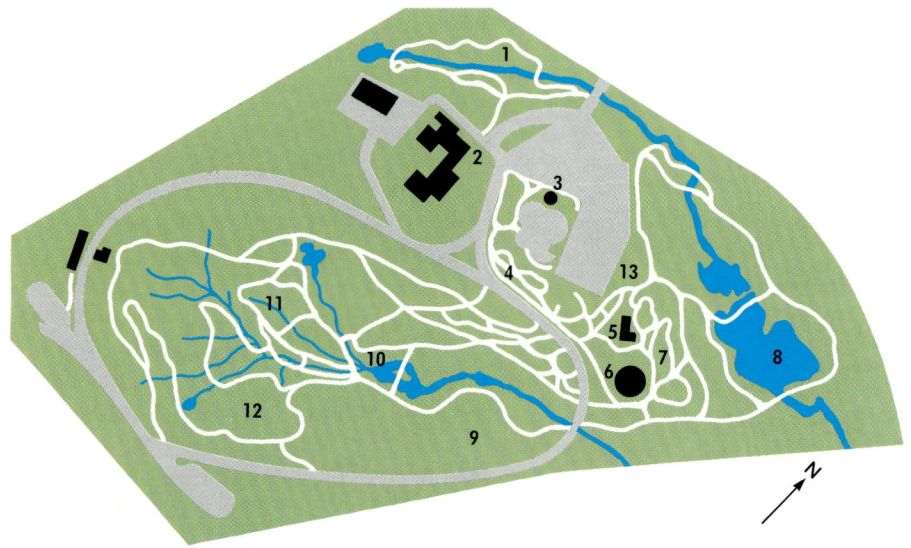

Left: The Geodesic Dome which houses tropical plants.

Left: Interior of the dome. The frame is fitted with acrylic panels.

Right: Ponds in the dome contain many waterlily species, including the *Victoria amazonica* from the Amazon.

The planting system throughout Stage II is also geographical. The soil in the area is clay and granite and is not very good, but over the years mulching and manuring have enriched it. With that accomplished, the curator of the Mount Coot-tha Gardens, Ross McKinnon, is confident that he has the best environment in Australia for growing an enormous range of flora. Said Harold Caulfield, the curator of the Gardens after Barry Dangerfield retired in 1980, 'By 1990 we might have one of the best Botanic Gardens in the world'.

Harold Caulfield retired in 1982, and Ross McKinnon was appointed curator of Mount Coot-tha Botanic Gardens the same year. Formerly in charge of the Brisbane City Gardens, McKinnon trained in the Adelaide Botanic Garden before coming to Brisbane.

The Mount Coot-tha Botanic Gardens offer many educational facilities, including a lending library of botanical books with about two thousand subscribers. There is a lecture hall, and also a laboratory for research. Like the Brisbane City Botanic Gardens they are administered by the Brisbane City Council. The Mount Coot-tha Botanic Gardens have a limited herbarium of exotic type specimens of plants contained within both botanic gardens. However the main herbarium for Queensland has, since 1968, been under the control of the Department of Primary Industries, and the library and collection of half a million herbarium specimens is now housed in the Department's building.

The first superintendent, Walter Hill, did not start a herbarium. He did, though, begin a botanical library upon which the library of today is based. In 1860 he sent £100 provided by the Queensland Government to Sir William Hooker for the purchase of books for Brisbane; many of them are rare and valuable today.

The present herbarium began with the plant collections made by Frederick Manson Bailey, who succeeded Walter Hill as colonial botanist in 1881. He collected plants throughout the state for thirty-four years, during which time he was transferred to the Queensland Museum. In 1902 there was a decision by the government to retire him and abolish the position, a decision reversed when Bailey refused to stop work whether he was paid or not, and upon public pressure to keep the position. He published in 1882 his *Synopsis of Queensland Flora* and three supplements, the *Queensland Flora 1899–1902*, and the *Comprehensive Catalogue of Queensland Plants*, 1913. His grandson C. T. White

Landscaping at Mount Coot-tha simulates natural terrain. Here a small river meanders through various plantings.

spent thirty-three years from 1917 onwards organising collecting expeditions in Australia, New Guinea, the Solomon Islands, the New Hebrides and New Caledonia.

The herbarium was enriched in 1882 by acquisition of part of the collection of plant specimens made by Robert Brown, the botanist with Flinders.

A copy of Brown's *Prodromus Florae Novae Hollandiae et Insulae Van Diemen* was purchased in 1861 and Flinders's charts are also in the collection of the botanical library.

Today Dr. R. W. Johnson is the government botanist and he organises a large staff supervising the work of taxonomists and ecologists in classifying native plants and general conservation work after mining activity. As in the earliest days of the colony, economic botany is also an important part of their work.

7

SOME REGIONAL GARDENS IN QUEENSLAND
ROCKHAMPTON
BOTANIC GARDENS

When Anna Pavlova, the great Russian ballerina, took her company of fifty dancers to Rockhampton in 1929, it is reported that she was much impressed with the Botanic Gardens.

Before and since her visit, tourists have been impressed with the vast Botanic Gardens on the western slopes of the Athelstane Ranges and on the edge of the Murray Lagoon. They rank, with the train that runs through the centre of the city, and the beautiful stone Customs House of 1890 on the Fitzroy River, as one of the distinguishing features of Rockhampton, and, as botanic gardens, are considered to rival those of Ballarat in Victoria as the best provincial gardens in Australia.

From the moment you approach and enter through a long avenue of 1930s bunya pines, *Araucaria bidwillii*, interspersed with bougainvillea, there is a sense of the Gardens being remarkable. Their size, today approximately 71 hectares, in relation to the population of 54 630 is impressive and, as you travel north, they are the first of Australia's botanic gardens in which you strike a tropical atmosphere.

An editor of the *Lachlander*, a Condobolin (NSW) newspaper, wrote in 1918:

I spent the most interesting half day in the Rockhampton Botanic Gardens, which in point of tropical growths exceed any gardens in the Commonwealth. I have emphasised tropical growths for it is in this respect that Rockhampton's Gardens stand unique. Yet within the limitation of its acreage and in all other all round features it will bear commendable comparison with the gardens of the Capital Cities of the Commonwealth.

The earliest part of the Gardens is the most interesting. It is known as the Lower Garden and is situated on the banks of the Murray Lagoon. Here are most of the one hundred species of palms grown in the Gardens, a spectacular cluster of twelve *Beaucarnea recurvata*, planted in the round, and a clumped European fan palm, *Chamaerops humilis*. Here, too, are most of the three hundred species of tropical trees from India, Burma, Thailand, Ceylon, tropical and South America, Africa, South-East Asia, the Pacific Islands and northern Australia. The most spectacular of these are the banyan trees, *Ficus bengh-*

alensis, the kapok tree from Ceylon, *Ceiba pentandra*; the silk cotton tree from the East Indies, *Bombax ceiba*; the sausage tree from Nigeria, *Kigelia pinnata*, with its metre-long seed pod shaped like a sausage; and *Schotia brachypetala*, the crimson flowers of which intoxicate the birds that flock upon it.

They are planted in the gardenesque style, as individual specimen trees or in groups designed to make a spectacular showing, and the professionally botanical nature of the planting is to be attributed to friendship and co-operation between the first curator of the Gardens, James Scott Edgar, and Ferdinand Mueller, the former director of the Melbourne Botanic Gardens. Edgar, appointed in 1873, had come from Kew Gardens in England. For Mueller, the tropical climate of Rockhampton and the availability there of a colleague with botanical knowledge provided a chance to experiment with the tropical flora he was sent from all over the world and which the climate of Melbourne could not support.

Rockhampton is situated on the Tropic of Capricorn and its average January maximum temperature of 31.4 degrees C, its July range of 8.6 degrees to 21.7 degrees, and its average annual rainfall of 856 mm mean that the Gardens are in the dry tropics. But the undulating site of the Gardens, with the possibility of irrigation from the Murray Lagoon and some waterholes on the land, provided conditions favourable to the constant lush growth of tropical species.

The Gardens also owe their excellence to a group of Rockhampton pioneers of scientific and botanical bent who fostered the Botanic Gardens idea and helped to establish them. Anthelme le Thozet was the first of them, and a botanist by profession. A Parisian who had been a member of the French Assembly – he was a Republican and a member of the extreme left – he had thought it advisable to emigrate from France after the coup d'état of 1851. In 1858 he migrated to Canoona, where gold had been found 10 miles north-west of Rockhampton, hoping to make his fortune as a publican on the goldfield. His canvas 'pub' was such a success that in a few years he sold it, bought a prime site for a second hotel in East Street, Rockhampton, and built the town's second hotel, the Alliance.

Above: *Ficus benghalensis*, the banyan tree, is one of many of the species which thrive in the Gardens.

Right: The main avenue of *Araucaria bidwillii*, the bunya pines which mark the entrance.

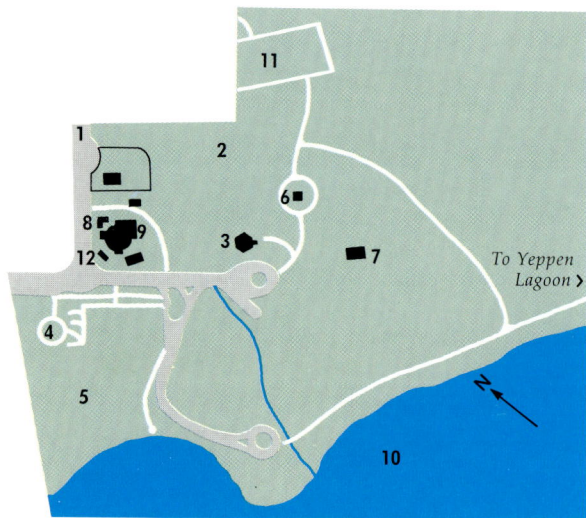

KEY TO MAP

1 Spencer St Entrance
2 Desert Garden
3 Kiosk
4 Aviary
5 Animal Section
6 Cenotaph
7 Picnic Shelter and Barbeque
8 Orchid House
9 Hugo Lassen Fernery
10 Murray Lagoon
11 Japanese Garden
12 Glasshouse

On arriving in the district he had written to Ferdinand Mueller in Melbourne offering to collect indigenous plant samples for him. The offer was accepted, and Mueller wrote in 1859 to the pioneers of Rockhampton, the brothers Colin and William Archer, who had started a farm a couple of miles from the incipient town, recommending 'Mr A. Thozet, a modest, intelligent and active man [who] is staying in your neighbourhood'. With the income from his second hotel, which he leased, Anthelme le Thozet built a house and large garden at Kalka, North Rockhampton, which he called Muellerville after Ferdinand Mueller. There his garden became 'one of the sights of Rockhampton', and he is credited with doing 'much for the Town of Rockhampton particularly in the matter of showing what could be grown there'. Crops of barley, oats, sugar cane, arrowroot, bananas, grapes, tobacco and wheat were planted. One stalk of wheat found floating in the river produced two hundred stalks, proving that with enough water wheat could grow well in Rockhampton. Le Thozet imported seeds of tropical trees and shrubs from France, and helped to train the young James Scott Edgar in the finer points of botany.

When the Municipality of Rockhampton was about to gazette some land for botanic gardens in 1869, it was Anthelme le Thozet who chose the 96 acre (38.8 ha) site for them, and plants from 'Thozet's Garden' were sent to the new botanic gardens in Townsville. He helped to establish the Rockhamp-

ton Gardens from his own imported stock of plants. He kept in touch with his professional colleagues in France and was awarded a medal from the Société Centrale d'Agriculture and Scientologie Générale for his discovery of the Australian fruit-boring moth; his book of research, *Roots, tubers, bulbs and fruits used as vegetable foods by Aboriginals of Northern Queensland*, was well received in Europe and he was responsible, though given no credit, for mounting the Queensland exhibit for the Paris Exhibition of 1867.

Besides Anthelme le Thozet and James Scott Edgar, the O'Shanesy brothers had an influence on the formation of the Botanic Gardens. Both had been trained as scientific gardeners. John O'Shanesy, from County Kerry, worked in the Brisbane Botanic Gardens when he arrived in Australia in 1861, and moving on to Rockhampton laid out a pleasure garden around the Ulster Arms Hotel for John McGregor. He began a nursery at Kabra, an orchard and a farm, and helped in a practical horticultural way to set out the Botanic Gardens. He collected plants for Mueller of the Melbourne Botanic Gardens. His brother Patrick also began collecting plants for Mueller, and became a Fellow of the Royal Society. He wrote gardening and botanical articles and collected wood samples that were exhibited at the Philadelphia Exhibition of 1876. He became a Fellow of the Linnaean Society of London, and later the Linnaean Society of New South Wales made him a member. He died young, but was influential also in the planting of the Botanic Gardens.

The site of the Gardens had originally been the Native Police Paddock, where police horses had been grazed and the soil, though poor, was loose and arable. The soil was enriched subsequently by ash from the bodies of goats that had overrun the town and been killed and burnt at the Gardens site. A beach of imported sand had been made on the banks of the Murray Lagoon, which became the main swimming spot for the Rockhampton population. One area had been a men's nude bathing pool, with the banks planted with clumps of bamboo for changing clothes in, and later a ladies' section was added. The dual purpose of the site, for Botanic Gardens and swimming pool, helped the popularity of the Gardens and by 1883, when the Corporation Baths were opened in Rockhampton, the Gardens were well established as a resort and pleasure park. In 1890 there was the benefit of a double-decker horse-drawn bus from the city to the Gardens; it was replaced in 1909 by a steam train along the same route. It was a 2 mile (3.2 km) trip from the town which, with the continuing discovery of gold and minerals, was 'a go-ahead place'.

As the first curator of the Gardens, James Scott Edgar had taken up residence in a rough slab cowshed converted for his use, and he began planting the 10 acres (4 ha) nearest the Lagoon – the Lower Garden. Not only did plant material come from Ferdinand Mueller in Melbourne, but Walter Hill, the superintendent of the Brisbane Botanic

*Above: **Licuala ramsayi**, a native Australian palm, in the Lower Garden.*

*Right: **Chamaerops humilis**, the European fan palm, is one of the famous sights of the Gardens.*

*Bottom right: A **Pandanus tectorius** cultivar.*

Gardens took a keen interest in the Rockhampton project, sending seeds and plants from the Brisbane Gardens. The huge banyan trees near the kiosk, the old palms, and most of the older araucarias and tropical economic trees of the Gardens date from this time.

Administered by a Botanic Gardens Trust, the Gardens were supported by a grant from the Queensland Government of £300 a year, and despite some floods, the capricious nature of the rainfall, and some cyclones, when James Scott Edgar retired in 1902 they were thriving as both a pleasure park and an economic botanic experimental garden. Many decorative trees, fruit trees and crops were acclimatised and tried before dispersal to the population of central Queensland.

Edgar's successor Richard Simmons, a Rockhampton nurseryman who had been trained in his native Ireland, became the curator of the Gardens for the next thirty years. He extended the planting to the higher level of ground, which became, in contrast to the dense planting along the edge of the lagoon, an open recreation park space known as the Upper Garden. Banyan trees were planted along the border between the Upper and Lower Gardens near the kiosk, which was built in 1912. The cenotaph . was built in 1924, and by then the Botanic Gardens were recognised as the main outdoor meeting-place for Rockhampton citizens.

The Hugo Lassen Fernery, built in 1929 of slatted timber as an elegant shadehouse, was a major addition to the Gardens. Named after its donor, a local dentist who loved the Gardens, today it holds a fine collection of tropical plants arranged in an attractive landscape. The fernery contains the best collection of *Aglaonema*, a fleshy plant with subtly patterned leaves found as understorey in the Amazon rainforest, to be seen in Australia, with over fifty species exhibited.

Left: The spectacular clump of *Beaucarnea recurvata*.

Below: *Beaucarnea recurvata* clump planted in the round and seen from a distance.

The third part of the Gardens developed during the curatorship of Simmons. This included the zoological and aviary section, established in the Lower Garden. The Murray Lagoon, with its wealth of local bird species, had already been declared a sanctuary in 1902, and in 1913 the Yeppen Lagoon also became a sanctuary. The zoological and aviary collection was enhanced in 1975 by the building of a flight aviary, which is today one of the important structural features of the Gardens, with over a hundred species of birds contained in it.

Up until the end of World War I, the Botanic Gardens were supported financially by a grant of money from the Queensland Government of between £300 and £500 annually, and extra money was raised through sales of plants grown in the Botanic Gardens Nurseries, occasionally reaching £1400 a year. Fruit trees and annual plants were the most popular sales. Subscriptions to the Gardens were sold for £400 a year, in return for which subscribers had access to free plants from the Gardens.

In the mid-1920s the Gardens are recorded as having a beautiful display of roses, zinnias, bougainvilleas, hydrangeas, penstemons, chrysanthemums, cinerarias, gladioli, lilies and hippeastrums. The croton display was said to be the best in the Commonwealth.

During the 1920s and 1930s the government subsidy diminished, and in 1930 the grant was £66 for the year. Plant sales, which had previously served much of the central Queensland district, dropped because of the growth of private commercial nurseries, and since the increased use of private motor cars enabled the Rockhampton population to go further afield than the Gardens for a day's outing and picnic, attendances at the Gardens dropped. By then Richard Simmons had retired. His son George Simmons became curator in 1931 and took charge of the Gardens for the next twenty-five years. He filled in the waterholes in the Upper Garden, and in 1935 began the pinetum, which today holds approximately fifty species of conifers, mainly from the genera *Cupressus* and *Chamaecyparis*. He had planted an *Araucaria cunninghamii* avenue interspersed with bougainvillea in 1930. The Palm Grove, with many of the coconut palms still there today, was planted near the archway covered with bougainvillea, which was also begun in the late 1930s.

Like most other botanic gardens, Rockhampton's suffered the loss of many valuable small plants during World War II through lack of finance for maintenance and adequate labour, although much effort was put into the growing of vegetables in the Gardens to provide food for the army divisions camped near Rockhampton. It was not until 1953 that they were flourishing again and a new entrance, known as the King George V Memorial Gates, was constructed at Spencer Street.

A new curator in 1957, K. C. Baker, was a horticulturist from Brisbane who developed further planting in the Upper Garden, and today his grove of kauris from Queensland is one of the sights of

Left: Calliandra haematocephala.

Below left: Schotia brachypetala, a tropical tree in the Gardens which produces flowers that intoxicate birds.

Right: Some of the Gardens' many croton species.

Far right: A collection of *Aglaonema* plants in the Hugo Lassen Fernery.

the Gardens. The fifth curator for the Rockhampton Botanic Gardens, Tom Wyatt, who came as a horticulturist from the Queens Gardens, Townsville, in 1974, is today engaged in projects for which there is a budget from the City Council of $100 000 a month.

The recreational, educational and horticultural functions of the Botanic Gardens have been improved, and there is potential to develop a further 72 hectares administered by the Gardens.

With the development of the Iwasaki Sangyo Queensland International Tourist Resort at Yeppoon, half an hour's drive from Rockhampton on the coast, there is expected to be an interest in Japanese culture and an influx of Japanese tourists to the Gardens. A Japanese garden has been built in the upper section of the Upper Garden of the Rockhampton Botanic Gardens. It was designed by Kenzo Ogata, one of Japan's best-known landscape architects, and constructed in 1981. Four Japanese gardeners, including Kenzo Ogata, spent ten months constructing it. For the traditional rocks they used 40 tonnes of blue granite from a quarry near the coast. Selected carefully for their shapes, some sentinel rocks are 2 metres high. Thirty fully grown trees growing in other parts of the Botanic Gardens were selected by the Japanese gardeners as being suitable for the new garden and were transplanted, including *Lagerstroemia speciosa* and the conifers *Podocarpus* and *Cupressus. Cupaniopsis*, trees native to the Yeppoon coast, were also chosen and transplanted. They are prized for their wind-shaped, gnarled habit.

The garden should take from five to ten years to develop. There is a lake and a tea house, and Japanese furnishings such as stone lanterns have

been placed in the garden. Plants are a mixture of native and exotic, with abelias, azaleas and pine trees all shaped by constant clipping. A high-maintenance garden, it is tended by Botanic Gardens staff; occasionally professional Japanese advice is given from the Iwasaki Resort. Though a distinct entity, the garden blends with the open parkland of the Upper Garden on its undulating site, and is visually part of the whole Botanic Gardens; the transition from Upper Garden Australian to Japanese is gradual, with the latter covering an area of 0.6 hectares.

An economic section of 32.3 hectares on land reclaimed from the Murray Lagoon in the Lower Garden is under development and will display timber and tropical fruit trees suitable for the Rockhampton climate. Like le Thozet's garden it will show what the area can grow. The aviary and zoological section has been extended, with an emphasis on the indigenous fauna that can be seen in the Rockhampton area. In the Upper Garden the area round the cenotaph, always a bit bare, is to be dressed with a ring of the Gallipoli pine, *Pinus pinea*, and on the drawing board at the moment is a Biblical Garden featuring all the plants mentioned in the Bible, a garden that will also be sited in the Upper Garden.

As an annexe to the Botanic Gardens, a new Australian Native Garden was begun on 40.4 hectares of reclaimed barren land on the outskirts of the northern end of the city in 1974. Already this holds three thousand Australian plant species, labelled as they are planted, and arranged according to habitat. The philosophy is to concentrate on flora of the northern half of Australia, flora that the Rockhampton climate easily supports, and plants from

Above: View of the Murray Lagoon from the Botanic Gardens.

the rainforest area, the sclerophyll supporting area and the deserts of the north are thriving. Two-thirds of the site, which, though originally flat, has been shaped to provide microclimates and to enhance the naturalness of the terrain, is irrigated. There are no steps, which will facilitate walking for the less agile visitors. This is mainly a recreational area, and there are barbecues and a slab hut, billy teas and damper, so that the public can recreate the pioneering days of gold and cattle in the Rockhampton area.

The scientific work of the Gardens is done by botanists in Brisbane at the Botany Branch and the Queensland Herbarium in the Department of Primary Industries under the direction of Dr. R. W. Johnson. With the additional new site, the Gardens have taken on an added importance, particularly with the concentration on growing indigenous species. The original Gardens, with their rare collection of tropical and subtropical trees, are a continuing delight.

BOTANIC GARDENS IN TOWNSVILLE

Though less impressive botanically than the botanic gardens of Rockhampton or Cairns, the Queens Gardens at Townsville, laid out on a formal plan, have a distinctive Victorian colonial charm.

With an annual rainfall of 1204 mm, a January temperature range of 23.8 degrees C to 31.3 degrees, a July range of 13.1 degrees to 24.9 degrees, and a pattern of wet summers and dry winters, Townsville's climate is typical of the dry tropics. Rainforest trees originating from all over the tropical world fill the Gardens, forming an oasis of shade. In one of the most arid, rocky landscapes of any large Australian city, the 4 hectare patch of dense green of the Gardens is vivid.

They are situated under Castle Hill, the dramatic red knob that dominates the town, and were proclaimed a botanic gardens reserve in 1870. A hundred acres (40.5 ha) of land close to the incipient town was set aside. In 1878, the first curator of the gardens, William Anderson, was appointed at a salary of £135 a year. A house was built for him, and on his recommendation an acre (0.4 ha) of grape vines was planted. Today, with the area shrunk to 4 hectares, allowing limited space for showing botanical species, a second garden of 28 hectares at Anderson Park has become the official Botanic Gardens, containing the nursery, some seven hundred plant species, and large shadehouses.

The Queens Gardens, adjacent to the playing areas of Queens Park, also have some botanical collections. Planting is organised on both an ecological and generic system. On the perimeter are straight avenues of tropical rainforest trees (dating from the nineteenth century), including many North Queensland species. There is an avenue of the black bean, *Castanospermum australe*, and mixed plantings of the African tulip tree, *Spathodea campanulata*; the mahogany tree from the West Indies, *Swietenia mahagoni*; the Indian almond tree, *Sterculia foetida*; the milky pine, *Alstonia scholaris*; the forest almond, *Terminalia catappa*; the beauty leaf tree, *Calophyllum inophyllum*; the mango tree, *Mangifera indica*; and *Ficus virens* and *F. benjamina*.

Some old and enormous banyan trees, *Ficus benghalensis*, are at the centre of the Gardens (some of these trees are thought to date from the 1880s). On one side of the Gardens boundary is the Rainforest Walk, with 166 species of labelled rainforest trees, shrubs and groundcovers from fifty-eight families and 122 genera, planted in simulated rainforest. A formal entrance is lined with a straight avenue of golden cane palms, *Chrysalidocarpus lutescens*, and lawns are embedded with bright annuals. On either side of the main axis is a maze of hedging plants, formally-clipped – plumbago, bougainvillea, acalypha, justicia – framing a small pool that terminates at the main entrance axis.

There is a sunken formal rose garden on the site of an old tennis court, and one pathway is covered with a large pergola on which grows a variety of tropical creepers: *Congea tomentosa*; the bleeding hearts creeper from West Africa, *Clerodendrum thomsonae*; *Thunbergia laurifolia* and *T. mysorensis*; *Clytostoma callistegioides* with delicate lilac and yellow flowers, from Argentina; and *Ipomoea horsfalliae*, a red morning glory from the West Indies. This pergola surrounds a small zoological section in the Gardens. Borders of cactus, a section of palms, and of pandanus lead the visitor onwards through the Gardens.

With the exception of the large banyan trees and the original trees planted round the perimeter of the Gardens, the planting and the design are all recent, a plan for a complete overhaul and redesigning of the Gardens having been called for by the Townsville City Council in 1950. The director of the Brisbane City Parks Department of the time, Harry Oakman, was asked his advice, and he chose his assistant landscape architect, Alan Wilson, to design the new Queens Gardens. The design was so successful that Alan Wilson was offered the position of superintendent of Parks and Reserves for Townsville, and he supervised the construction of the new Gardens when he took up the position.

Later he designed the Botanic Gardens at Anderson Park in a classical picturesque landscaping style, incorporating lakes set in rolling countryside. On a site that was once market gardens and sugar cane fields, the Botanic Gardens are now in the centre of the developed area of Townsville on the Ross River Road en route to Charters Towers. Planting began in 1962.

The earliest trees on the land were planted in 1932 on 7 acres (2.8 ha) of land acquired for Anderson Park, named for the first curator of the Botanic Gardens. Two hundred trees were planted and today these, mainly *Samanea saman* trees planted in clumps, and *Melaleuca* and *Eucalyptus* species indigenous to the area, dominate the landscape.

Many specimen trees have been planted either singly or in generic groups in the Botanic Gardens area, which was acquired after the park land. A large collection of palms and a collection of pandanus, mainly indigenous plants from the Weipa iron ore ranges and consisting of forty to fifty species, were planted. The pandanus collection, gathered by Robert Tucker, a horticultural employee of the Gardens, is believed to be the largest in Australia. Today the *Samanea saman* trees at the main entrance form a dense shade canopy and the early plantings of *Livistona*, *Sabal* and *Washington* palms make an impressive park landscape. The lagoons attract indigenous birdlife. As Townsville's rainfall occurs mainly in summer, irrigation is necessary during winter months and numerous wells on the land and sprinkler systems keep the growth lush. Planting work is ongoing, with an extensive nursery developed, and the Gardens will not be finished for many years.

This late development of Townsville's botanic

Top right: Open parkland at Anderson Park contains many species of tropical trees.

Right: The Queen's Gardens in Townsville is dominated by Castle Hill.

KEY TO MAP OF THE BOTANIC GARDENS ANDERSON PARK

1	Meliaceae	16 Lakes
2	Anacardiaceae	17 Felicinae
3	Elaeocarpaceae	18 Musaceae
4	Small Trees and	19 Cycadaceae
	Shrubs in Family	20 Palmaceae
	Groups	21 Ulmaceae
5	Bush Houses	22 Gramineae
6	Economic and	23 Sapotaceae
	Medicinal Plants	24 Apocynaceae
7	Residence	25 Verbeniaceae
8	Nursery	26 Myrtaceae
9	Rose Garden	27 Bignoniaceae
10	Guttiferae	28 Proteaceae
11	Flacourtiaceae	29 Salicaceae
12	Lecythidaceae	30 Casuarinaceae
13	Sterculiaceae	31 Coniferales
14	Rainforest Plants	32 Fountain Garden
15	Succulents	

gardens atones for an almost complete neglect of the original 100 acres in the centre of the town set aside as a botanic gardens reserve in 1870. Although in 1878 the Council advertised for a gardener to lay out the gardens and maintain them, and William Anderson was appointed curator and practical gardener, the Council resolved that all ordering of plants be left in the hands of the mayor.

By 1884 the Gardens had been removed from the control of the Council and were being governed by four trustees. Eighty-four acres (34 ha) of them had been given to provide land for the gaol, the lunatic asylum, the grammar school and some playing fields. The Botanic Gardens remained 'very little laid out', according to a report of the day. In November 1884 the Trustees advised the Council that they considered the land unsuitable for a botanic garden and proposed to dispose of the site and acquire another area. This, however, did not happen, although it appears that the Botanic Gardens were not outstandingly lovely. According to Townsville historian Dorothy Gibson-Wilde, the exceptions to the general rule of barrenness in Townsville were the Belgian Gardens and the Hermit Park/Mundingburrah regions, where the estates of the rich boasted colourful gardens with fine trees that, with Gulliver's Nursery, 'put the Botanical Gardens to shame'.

Some trees in the Gardens were planted by the Queensland Acclimatisation Society. Trees were also sent from Anthelme le Thozet's fine garden of exotic trees and shrubs in Rockhampton. Others were sent from Brisbane by Walter Hill, the curator of the Brisbane Botanic Gardens, who had visited Townsville in 1873.

By 1894, after various disruptions in management during which the Council requested the trustees to resign and resumed control of the Gardens, William Anderson's salary as curator was reduced to £120 per year. As funds were short, it was recommended that the curator grow only sufficient flowers to decorate the gardens. Young trees grown there were sold to Rockhampton's Homestead School. Charges were made to help defray costs, with the hire of pot-plants raised at the Gardens costing 10s 6d per day, the use of the Gardens for picnics 10s 6d per day; for musical entertainments there the fee was £1 1s10d per day. Though lacking in botanical excellence, the Gardens were extremely popular as a pleasure ground for the Townsville population. In 1890 the Council records note that the Rose Bud Club held a Gypsy Tea in Queens Gardens, and many similar functions must be presumed to have been held there. William Anderson remained the curator until his retirement in 1900.

From the turn of the century until their remodelling in 1950, the Queens Gardens were maintained as a park rather than a botanical garden. During World War II, when Townsville became a military base for approximately 100 000 American troops, the park became a camp for deserter soldiers, and of the park-planting only the nineteenth-century

Opposite: A hedge of the climber *Thunbergia laurifolia* in the Queen's Gardens. — *Inset: Thunbergia laurifolia,* close up.

Above top: Close up of *Mucuna bennettii,* a creeper from New Guinea, in the shade house at Anderson Park.

Above: A hybrid of *Anthurium andraeanum* at Anderson Park.

Left: **The main axis of the Queen's Gardens has formal flower beds.**

Bottom left: **Young trees at Anderson Park will take many years to mature.**

Above: **Rainforest in the Queen's Gardens contains many tropical species.**

Right: ***Samanea saman*** **trees at Anderson Park are the oldest plantings.**

trees survived. But over the years many of the original plants sent by le Thozet and Walter Hill have disappeared.

Cyclones severely damaged the Queens Gardens in 1896, in 1903 and in 1971, when Cyclone Althea at Christmas wrought such havoc that a hundred truckloads of debris had to be removed from the Gardens. The superintendent of Parks, Jim Thomas, who succeeded Alan Wilson in 1968, saved thousands of trees uprooted in the winds by replanting them. He also engaged in drastic tree surgery. (He later provided help after Cyclone Tracy in Darwin in 1974.)

Formerly superintendent of the Royal Botanic Gardens in Sydney, Jim Thomas was a trainee for many years at the Royal Botanic Gardens, Kew, and his policy is to maintain the redesigned Queens Gardens and finish the new Gardens at Anderson Park. He built the rosery in the Queens Gardens and substantially rehabilitated them after Cyclone Althea. The herbarium and scientific research and botanical work in Townsville is carried out at the James Cook University, the CSIRO Research Laboratories, and the Department of Industry in Townsville as well as at the Queensland Herbarium. Throughout Townsville, thousands of street trees have been planted, and the general appearance of the city, although still dry, is remarkable compared with its early arid days. The Gardens are for the first time being recognised as an important Australian botanic garden. The tropical trees of Queens Park are particularly treasured, and a Bicentennial grant of approximately $120 000 has been made for the creation of a palm collection. The new Palmetum, situated on approximately 20 hectares of land on the Ross River, five kilometres from Townsville, holds approximately 300 palm species including most of the Australian palms. This annexe of the Botanic Gardens at Anderson Park, promises to be one of the finest palm collections in Australia, as the dry Townsville winters increase the range of palms able to be sustained.

BOTANIC GARDENS IN CAIRNS

Visit the Flecker Botanic Gardens in August and you will see near the entrance the shower of orchids vine from Thailand, *Congea tomentosa*, covering a mangosteen tree, *Garcinia mangostana*, in a veil of dusky mauve. It is one of the breathtaking sights offered by Australian botanic gardens.

The *Congea* and other tropical flora from throughout the world perform better here than in any other Australian city, for Cairns, 17 degrees south of the equator and in the shadow of the Great Dividing Range, is the only large city in the wet tropics. With an average annual rainfall of 2001 mm, it is one of the wettest places in the country, with 85 per cent of the rain falling between December and March. Here plants like heliconias, stephanotis, rare tropical orchids and philodendrons, which in the dry tropics need irrigation to thrive and in the southern states need glasshouses, can grow in the open air throughout the year. The temperature, with a January mean maximum of 31.5 degrees C and minimum of 27.5 degrees C, and a July mean maximum of 23.6 degrees C and minimum of 16.7 degrees C, is the other necessary ingredient for such profuse growth.

It is a climate that has produced, in an 80 kilometre radius of Cairns, what is believed to be one of the richest resources of indigenous flora and fauna to be found in the world. The Botanic Gardens complex, which has grown to 319 hectares of public recreation space involving many types of ecological environments (including fourteen types of rainforest), demonstrates the prolific indigenous plant life.

Three main sections of the Botanic Gardens complex adjoin each other at Edge Hill, 4.8 kilometres from Cairns and just off the Captain Cook Highway. The earliest and most formal part of the Gardens is the Flecker Botanic Gardens, begun in 1881. These Gardens, of approximately 3.25 hectares, are designed in an informal Victorian style and comprise grassy patches, ordered paths, and groups of

---

Left: Mucuna bennettii, flame of the forest vine, grows over trees in the Flecker Botanic Gardens.

large old exotic tropical trees underplanted with shrubberies; there are bush-houses and a nursery.

In the centre, near the office, is a *Samanea saman* tree with a spread of 30 metres and a trunk 9.7 metres in circumference, which stands where the original nursery was. Round this is a group of tropical crop plants and medicinal plants, many of which are early plantings and include coffee, tea, cinnamon, nutmeg, turmeric, a mangosteen, *Garcinia mangostana*, and from Burma, the *Hydrocarpus* tree from which chaulmoogra oil is taken: this is used to treat leprosy. These with other fruit trees in the Gardens form a collection of approximately twenty-two types of economic trees.

Paths radiate from this section, one leading to a rainforest area where tall trees are underplanted with a ground covering of marantas, heliconias, spathyphyllums, gingers, aspidistras and philodendrons. This leads to a gully built round one of the two creeks running through the Gardens, which flows for nine months of the year, heavily shaded by the large and rare rainforest canopy above. Palms, including a 10 metre high *Caryota* palm, are thickly planted as an understorey, and *Dracaena fragrans* with its wonderful night scent is at ground level as are heliconias and gingers.

The walk from here is designed in a figure eight. On a lawn of broad-leaved buffalo grass, *Axonopus compressus*, the predominant lawn grass in the Gardens, are many of the older flowering tropical trees, among them the pudding pipe tree, *Cassia fistula*. Other plants are *Tabebuia riparia*, with its purple flowers; the giant crepe myrtle from India, *Lagerstroemia speciosa*; a eugenia hung with the flame of the forest climber from New Guinea, *Mucuna bennettii*; bauhinias, caesalpinias, parkias, and the climbers *Thunbergia mysorensis* and *Faradaya splendida*. Here also you can see the native Cairns bignonia, *Neosepicaea jucunda*. The best time for seeing the flowering trees is in November and December, but if you visit in June, July or August you will see the *Mucuna* and *Congea* climbers flowering.

Other trees on the lawn have curiosity value, like the Leichhardt's pine, *Nauclea orientalis*; *Barringtonia asiatica*, a native of the Cairns coast, where its seeds, which are poisonous, were used by Aboriginals to stun fish; and the Queensland tar tree, *Semecarpus australiensis*, the sap of which can cause burns. There is a good example of the Spanish lime, *Lophostemon lactisluus*, and the cannon ball tree, *Couroupita guianensis*.

Across the small bridge you come to another lawn with a predominant *Peltophorum pterocarpum*, a large tree with bright yellow flowers, and a fine example of the teak tree from Burma, *Tectona grandis*, with large green leaves and panicles of white flowers. Along the front of Collins Street the thick planting of trees forming the boundary to the street includes *Stenocarpus* and poinciana, which add to the range of flowering trees in the Gardens.

A new orchid house, situated in the Munro-Martyn fernery, holds the Gardens' collection of tropical orchid species, with *Dendrobium*, *Cattleya*, *Phalaenopsis*, and *Oncidium* being well represented.

The Flecker Botanic Gardens today also hold over two hundred species of palm, one of the most representative palm collections in Australia. These palms are scattered throughout the whole complex, according to their varying habitat requirements and seeds from them have been exported to other botanic gardens within Australia and overseas.

Because space was so limited at the Flecker Botanic Gardens, the Gardens spread in 1976 across Collins Avenue to what was once swamp land. Today this area of 28 hectares holds the Centenary Lakes, which commemorate the centenary of the 1876 founding of Cairns. Freshwater Lake consists of 3 hectares and Saltwater Lake approximately 1.3 hectares. There are melaleuca wetlands, mangrove wetlands, and Palm Swamp, and some fine open grassy spaces landscaped with clumps of indigenous trees and palms.

The most outstanding feature is the long boardwalk, known as the Jungle Walk, winding through thick melaleuca and pandanus forest about a metre above the swampy, spongy ground. Some of the melaleucas have trunks up to 7 metres in diameter, and there are intermingled pandanus, lawyer vine or *Calamus*, and palms, particularly the Alexander palm, *Archontophoenix alexandrae*. Epiphytes are naturalised in the upper branches of the trees and the next project proposed for the lakes is to construct a more elevated boardwalk so that life at treetop level can be observed. This is typical of the type of forest that surrounded Cairns when the first settlers came. It contains many of the 105 species of birds observed in the Botanic Gardens, and some brilliant butterflies, which are more diverse and spectacular here than anywhere else in Australia.

In the open grasslands going towards the Freshwater Lake, you pass clumps of melaleucas on the higher sandy ridges. The lakes have five *Melaleuca* species: *M. quinquenervia*, *M. viridiflora*, *M. dealbata*, *M. leucadendron* and *M. cajuputi*. Various kinds of pandanus are also sculpted into the landscape, as are eucalypts and palms. A notable transplant is the *Cycas media*, near the Freshwater Lake, a thousand-

Above top: *Cycas media*, probably a thousand years old, transplanted from the bushland to the Centenary Lakes.

Above: Rainforest walk in the Flecker Botanic Gardens.

Right: *Congea tomentosa*, the shower of orchids vine, behind a *Livistona chinensis*, the Chinese fan palm, near the entrance to the Flecker Botanic Garden.

year-old cycad brought from the Mount Lumley area of the Gardens.

With the formation of the lakes and the subsequent consolidation of land produced by their excavation, formerly unusable swampland has become valuable. The Freshwater Lake, with a depth of about a metre, is covered with indigenous *Nymphaea* and occasionally houses crocodiles. Its construction has helped to eradicate the anopheles mosquito, as the waters can now be treated to prevent their breeding. During World War II, troops stationed in the Cairns area were infected with malaria from the mosquito. A collection of palms including a rare *Nypa fruticans, Orbignya cohune, Neodypsis decaryi, Corypha elata* and *Hydriastele wendlandiana* have been planted round the lake. The Saltwater Lake, which is tidal, is bounded by mangrove forest and accompanying wetlands flora. Between the lakes is open parkland with areas defined for recreation (including barbecues).

The third part of the Gardens, which adjoins the Flecker Botanic Gardens, is a hillside section of forest leading to the peaks of Mount Whitfield, 370 metres high and Mount Lumley, 304 metres. The forest up to 300 metres is open dry sclerophyll forest, and after that height the tropical rainforest takes

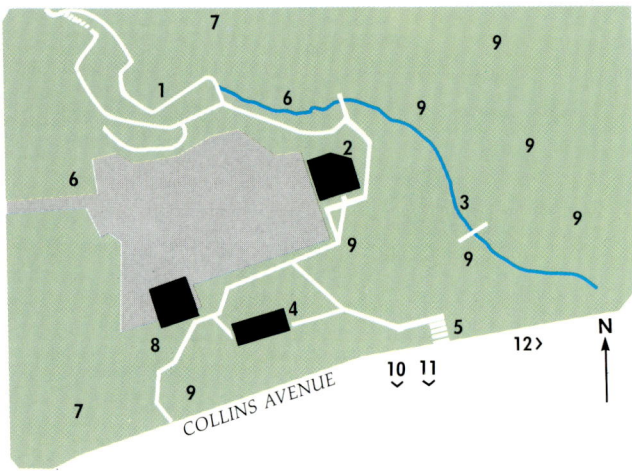

over. In this part of the Gardens are 9.6 kilometres of walking tracks, made in the 1970s under the Regional Employment Development Scheme, which wind through the reserve with its deep gullies and often steep hillsides. One walk takes about an hour, and a longer extension of this can take four hours.

On the way you pass approximately eighty-nine species of tropical trees, many of which are named; all are listed in a pamphlet issued by the Gardens. Here you can see the famous rainforest timber trees of northern Queensland: the black bean tree, *Castanospermum australe*; the milky pine, *Alstonia scholaris*; the red siris, *Paraserianthus toona*; the scrub turpentine, *Canarium muelleri*; *Palaquium galactoxylum*, and the Burdekin plum tree, *Pleigynium timorense*. Also here is the birdlime tree native to the coral cays of the Barrier Reef, *Pisonia umbellifera*. The lower storey of the forest is of palms, particularly the Alexander palm, *Archontophoenix alexandrae*; tree ferns, *Cyathea cooperi*; and melaleucas. Threaded through this section are three genera of the family Cycadaceae, including five species of *Lepidozamia*.

There are also many exotic tropical trees and shrubs throughout the walks, the most lovely of which is the large section of bamboo forest about halfway up the hillside, which has been there since at least 1909, when early photographs were taken of it. A kapok tree from Ceylon, *Ceiba pentandra* (from which kapok for bedding comes), the African tulip tree, *Spathodea campanulata*, and the poinciana, *Delonix regia*, are also to be seen here. These were planted by the founder of the Gardens, Eugene Fitzalan, who had his house up on the hillside above what are now the Flecker Botanic Gardens which, being almost below sea level, were prone to flooding. A renowned lover of tropical orchids and a collector of indigenous rainforest species, he settled in the Edge Hill district in 1881 with his family in a house called Rosebank, which no longer exists. He is believed to have grown the first rose to bloom in the Cairns area.

According to early historical reports, a visit to Fitzalan's Botanic Gardens on the slopes and flat of Edge Hill was a popular day out of town for Cairns citizens of the 1880s, 1890s and early 1900s, particularly after the beginning of the railway from Cairns to Edge Hill in 1885. All the land now included in the Botanic Gardens complex was gazetted by the Cairns Council for recreation in 1886. Eugene Fitzalan's Botanic Gardens were ceded to him so that he could sell his plants, particularly native species he had collected and propagated in his nursery and was sending throughout the country, in exchange for his opening his gardens to the public.

An Irishman, Fitzalan had been orphaned at the age of twelve, when his father, an architect, died. He was then living at Enniskillen, where he was being educated at a private school for a career in medicine. (Botany was one of the subjects he studied.) Having to support himself, he took work in the large gardens of the local rich. After four years of this he travelled for another two, working at one time for the London nursery of J. H. Veitch, which was well known throughout England. In 1849 he

migrated to Melbourne, where he met Ferdinand Mueller and began collecting indigenous specimens for him in his spare time (he was engaged in prospecting on the Victorian goldfields; then, working in Geelong, designed and planted out a 30 acre (12.1 ha) landscape for a Mr Thorn). He attempted to start a botanic garden in Geelong and by 1860 was being referred to by Mueller in reports of the Melbourne Botanic Gardens as 'our' collector.

In the mid-1850s he went to live in Brisbane, where he started a seed and nursery business. In 1860 he became the official botanist on the expedition of the *Spitfire* in northern waters to discover the mouth of the Burdekin River, sending his findings on the local botany back to Mueller, who published them. He lived in Bowen, in northern Queensland, as government contractor for twenty years, collecting native species in his spare time. After arriving in Cairns and starting his nursery, he collected assiduously, and became known throughout Australia as an official collector of orchids from the wild. Two orchids, *Eria fitzalanii* and *Eulophia fitzalanii*, were named after him, as was *Randia fitzalanii*, a deliciously perfumed flowering tree which grows profusely on Magnetic Island off Townsville.

Fitzalan's Gardens in Cairns were the basis for the Flecker Botanic Gardens of today. The modern design is the work of Vince Winkel, the curator and director of the Parks and Gardens Department of the Cairns City Council from 1966 to 1984. The Gardens, when he inherited them, were rundown. After Eugene Fitzalan left, they were cared for by gardeners until 1923, when L. Wright, a nurseryman from Townsville, took charge of them. An expert on palms and butterflies, he maintained the gardens and established a small zoo, which when he retired in 1947 was abolished. T. Mitchell and Jim Gould, who built some lawn areas in the Gardens, were the successive curators until the appointment of Vince Winkel.

He cleared the area round the office of lantana and weeds to 'find' the early plantings of economic trees – some of the lawns had guinea grass 3 metres high. The creek through which the Rainforest Walk was built was a mere swampy trickle full of rocks, which was subsequently cleared and formed into the running stream it is today. He oversaw the construction of a new fernery, and improved the collections of plants, particularly the bromeliads and the crotons, which numbered three hundred species. He was particularly keen to collect palms, increasing

Left: **Freshwater Lake in the Centenary Lakes area of the Botanic Gardens.**

Top right: **Steps down to the creek bed in the rainforest walk at the Flecker Botanic Gardens.**

Right: **A stand of *Archontophoenix alexandrae*, Alexandra palms, native to the Cairns area, in the Flecker Botanic Gardens.**

the collection to over two hundred species, many of which were planted in pots.

His landscaping attention was focused on the need for pleasing the visitor through varying the leaf textures, creating anticipation by having curving pathways leading round corners. He believed in forming groups of trees and disliked specimen planting, and when the project for the Centenary Lakes was suggested by the Cairns City Council, he was able to express his style on a new canvas, and on a large scale.

Under this curatorship, the old gardens became official Botanic Gardens rather than a municipal pleasure park and were designated the Flecker Botanic Gardens in 1971, in honour of the contribution to them by Dr Hugo Flecker. A radiologist, he had come to Cairns in 1932 and became interested in natural history, particularly the toxic qualities in the flora and fauna surrounding Cairns. He worked on the Queensland finger cherry and tar trees, and following two deaths from the marine box jellyfish, he identified *Chrionex fleckeri*, the deadly box jellyfish of northern waters, named after him. He founded the North Queensland Naturalists' Club, and in 1958 the members decided to form a native preservation society and to build a botanic garden. They contributed to its gradual evolution from pub-

lic recreation space to more scientifically based Botanic Gardens. Hugo Flecker collected herbarium samples from the wild, and his collection is now incorporated into the CSIRO Herbarium in Atherton.

The subsequent director for the Parks and Recreation Department of the Cairns City Council, which controls the Botanic Gardens, was Robert Guthrie, a horticulturist and a landscape architect. He enriched the already outstanding collection of palm species and introduced a policy of uniform labelling of plants within the Gardens system. He also added a bush-house to the Munro-Martyn fernery building.

The Gardens are funded by ratepayers to the Council, and in the past finance for their development has been limited. The Council recently completed a three-year programme of development to bring the Gardens up to an international standard, with improved nursery facilities and the renovation of the office and visitor centre in what was once the curator's cottage in the grounds of the Flecker Botanic Gardens. The new director for Parks and Recreation for Cairns, James R. Malcolm, is a bachelor of science and landscape architecture graduate from New York State University and was previously the superintendent for planning and works for the Melbourne City Council. He aims to employ

Left: **Saltwater Lake in the Centenary Lakes area of the Botanic Gardens.**

Above: **A ground cover of *Alternanthera dentata* cultivar in the Mount Lumley and Mount Whitfield Reserve.**

Top right: **A stand of bamboo in the Mount Whitfield Reserve, probably planted by Eugene Fitzalan at the end of the 19th century.**

Right: **Pathway leading to the forest reserves of Mount Lumley and Mount Whitfield.**

a technically qualified botanical curator to devote attention exclusively to the Botanic Gardens and their interpretation by the public.

Appointed in July 1987, James Malcolm is working on a plan to link physically the three different sections of the Gardens complex and to provide a rainforest education centre within it. He is confident of the future of the Gardens in scientific terms as the Cairns climate presents the optimum conditions for growing rare tropical plants from all over the world.

8

THE BOTANIC GARDENS OF ADELAIDE

When the first flower of the *Victoria amazonica* waterlily bloomed in the Adelaide Botanic Garden in the late summer of 1868, daily newspapers gave an hour-by-hour description of its opening. Anyone who had any spare time hurried to the Garden to see the bud open, with 'delicately tinged petals of exotic form, reaching its best at 4.30 p.m. when it expanded its aura of pure white petals and exhaled a delicious perfume'.

The plant had been seen in 1837 in the Amazon rainforest by Dr Richard Schomburgk, later to become director of the Botanic Garden. Its leaves, spreading to 2 metres in diameter, were so large that a careful child could stand on them. The flowers were 33 cm across, lasting for only two days before sinking below the surface of the water to set seed. The Victoria Waterlily, in its heated Victoria House, became one of the sights of Adelaide: 30 000 visitors came to see it during its first five weeks of flowering. Victoria houses, some of which were designed with water wheels that emulated the sluggish waters of the Amazon, were fashionable throughout the botanical world from 1850 to 1890, and Adelaide had

Left: An Indian-style summerhouse against a background of ilex trees in the Adelaide Botanic Gardens.

and still has Australia's best. The Botanic Garden, in the late nineteenth century, was an enormous source of state pride.

Today South Australians are still proud of their botanic gardens, which have an international as well as a national role to play. The state now has three major botanic gardens within a radius of 26 kilometres of the city, and together they support a wider range of flora than is grown in any other state of Australia. The oldest of these gardens, 18.5 hectares in extent, was established in 1855 on North Terrace, 1.6 kilometres from the Adelaide GPO. Here the climate is Mediterranean, with hot, dry summers, an average annual rainfall of 533 mm, and a distinct change of temperature between winter and summer. The maximum temperature for July averages 14.8 degrees C and the average minimum is 7.9 degrees. The mean maximum temperature for January is 28.5 degrees C and the mean minimum 17.1 degrees. In the summer, when the hills round the city turn the colour of straw, the temperature can stay as high as 37 degrees C day and night for three or four days at a time. It is not hard to understand why some deciduous trees and shrubs have difficulty thriving on the Adelaide plains.

But at Mount Lofty, twenty minutes drive away in the hills, a second botanic garden was initiated

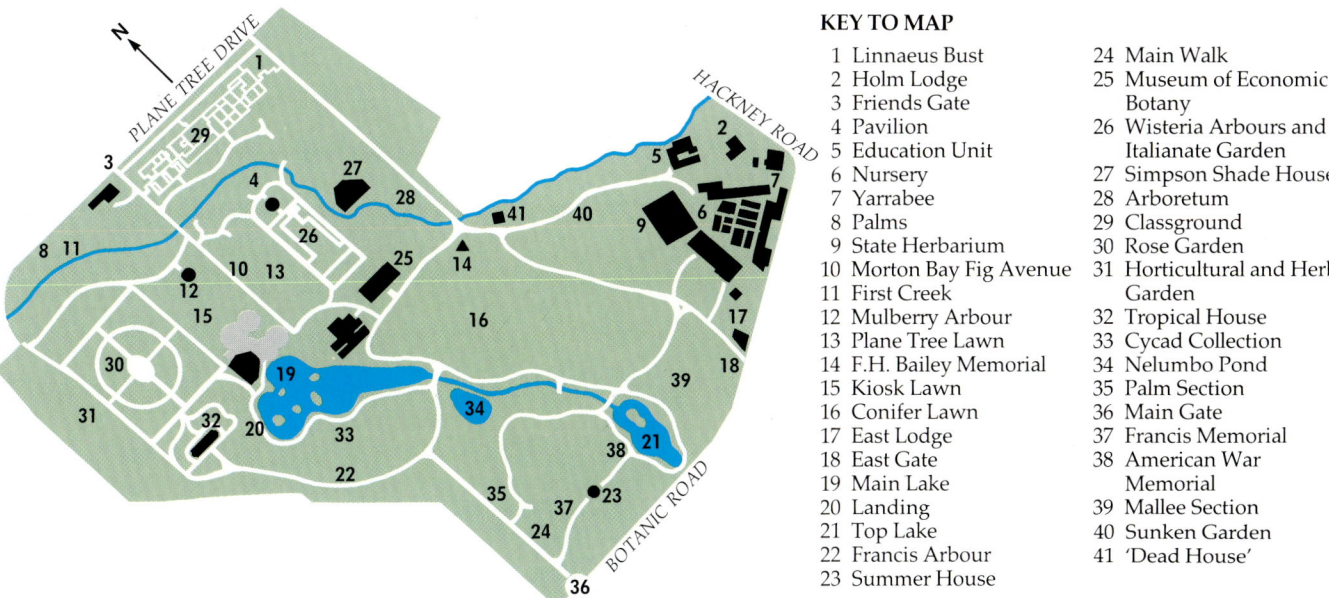

KEY TO MAP

1 Linnaeus Bust	24 Main Walk
2 Holm Lodge	25 Museum of Economic
3 Friends Gate	Botany
4 Pavilion	26 Wisteria Arbours and
5 Education Unit	Italianate Garden
6 Nursery	27 Simpson Shade House
7 Yarrabee	28 Arboretum
8 Palms	29 Classground
9 State Herbarium	30 Rose Garden
10 Morton Bay Fig Avenue	31 Horticultural and Herb
11 First Creek	Garden
12 Mulberry Arbour	32 Tropical House
13 Plane Tree Lawn	33 Cycad Collection
14 F.H. Bailey Memorial	34 Nelumbo Pond
15 Kiosk Lawn	35 Palm Section
16 Conifer Lawn	36 Main Gate
17 East Lodge	37 Francis Memorial
18 East Gate	38 American War
19 Main Lake	Memorial
20 Landing	39 Mallee Section
21 Top Lake	40 Sunken Garden
22 Francis Arbour	41 'Dead House'
23 Summer House	

Above: **The main gates of the Botanic Gardens came from England in 1880.**

Below: **Nineteenth century classical reproduction lead statues along Main Walk.**

Top right: **The Palm House, designed in Bremen and built in 1875.**

Right: **Main Walk of the Adelaide Botanic Gardens has a 19th century atmosphere with statuary, old trees and shrubbery beds.**

Opposite: **First Creek, a natural water course which runs through the Gardens.**

in 1952. It has a temperate climate, a rainfall of up to 1525 millimetres annually, and a supply of spring water. At night the temperature drops considerably – in summer it can fall from 40 degrees to 20 degrees. The soil at Mount Lofty is acid in contrast to the highly alkaline soil of Adelaide. Here plants from the cooler temperature zones of the northern and southern hemispheres can be grown easily, and there are glades of azaleas beneath exotic maples and birches. Also in the hills, at Blackwood, is 'Wittunga', a garden featuring the best collection of South African ericas outside of South Africa. It also has a collection of proteas and leucadendrons, banksias and a spectacular collection of Western and South Australian sandplain plants.

But when the Botanic Garden is spoken of in Adelaide, it is the city garden people are referring to. This still retains some of the nineteenth-century atmosphere it had when the *Victoria amazonica* waterlily first bloomed. It has the Palm House, im-

ported piece by piece from Bremen in 1875, and considered still to be the finest piece of glasshouse architecture in Australia. It has an august avenue of Moreton Bay figs, an avenue of araucarias, and one of London plane trees planted in the 1870s. The main gates were opened for the inauguration of the new road to the Garden in 1860, and subsequently for visits of Royalty, but are otherwise closed; they arrived in their present form from England in 1880. They and the statuary of the Garden have an elegance unusual in other Australian botanic gardens. The Garden still has Australia's only Museum of Economic Botany, built in 1880 and opened in 1881. And among the botanical curiosities is a lake dedicated to the sacred lotus, *Nelumbo nucifera*, the only

such lake in an Australian botanic garden. The Garden still has the *Victoria amazonica* waterlily and, until 1976, it was the only Australian botanic garden to grow it consistently. (There is one now in the Mount Coot-tha Botanic Gardens, Brisbane.)

The collection of plants from the Mallee and the dry inland areas of Western and South Australia is particularly fine. This is also the only botanic garden in Australia to have a systematic class ground displaying the classification of plants, with species arranged according to the system of Engler and Prantl in formal beds, and explanations provided in a pamphlet. The collection of *Eucalyptus* is interesting historically as it includes the first stand of native flora grown for public exposition in Australia; planted in the 1860s, the specimens are large and well formed.

The Garden, particularly its older section, is Victorian in style: a nineteenth-century series of 'linked rooms' with associated types of flora, leading from the gardenesque planting at the entrance.

It follows the line of First Creek, a watercourse meandering through the parkland between North Terrace and a tributary of the River Torrens. The creek becomes a series of connected lakes, cleaving the Garden into two distinct sections. (The northern section of the Garden appears much larger due to the inclusion of 29.5 hectares of Botanic Park.) The flat area is the oldest part of the Garden, and the undulating open land on the northern side of the creek houses the modern administration and herbarium buildings, set in a landscaped parkland.

The traditional approach to the Garden is through the main gates on North Terrace and up Main Walk, a straight axis past the white Molossian Hounds, replicas of the statues of Papal Lions guarding the Vatican. This and other cast-iron statuary made in Germany was removed from the Garden during World War I but reinstated later. You pass the Upper Lake, with its ornate brick paving and thick planting of shrubs and palms leading to a memorial dedicated to the Australian–American alliance during World War II, then walk past large trees planted during the 1860s: the hoop pine, *Araucaria cunninghamii*; the Norfolk Island pine, *Araucaria heterophylla*; and the Chilean wine palm, *Jubaea chilensis*, planted in 1901 by the Duke and Duchess of York. There is a particularly large old red cedar, *Toona australis*; a specimen of the South African tree *Schotia brachypetala*; and, on the skyline to the west of Top Lake, a Queensland kauri pine, *Agathis robusta*, that is over 33 metres in height – it is considered to be the finest tree in the Garden.

The Main Lake is near the Simpson Kiosk, a modernised restaurant that was formerly part of the Zoological Gardens. Where the lake widens is a spectacular collection of *Nolina* species and cycads. Collected from throughout the Garden, they are now grouped together as part of the Garden policy to rationalise the arrangement of collections made over the last century. The present director of the Botanic Garden, Dr Brian Morley, explains that 'we have the plants, it's a question of rearranging them to better advantage'. The policy of rationalisation has been continued through the 'rooms' of the Garden: the Plane Lawn with its fine tall plane trees; the nineteenth-century rosery with 250 rose varieties; a section of ground-covering plants; the class ground; the herb garden; a section of espaliered fruit trees; the Schomburgk Range glasshouses with displays of begonias, bromeliads, ferns and cacti; the Victoria House with its tropical aquatic plants; and a section showing outdoor terrestrial bromeliads.

Dr Morley is a taxonomist, formerly from the Royal Botanic Gardens, Kew, and Glasnevin, the National Botanic Garden of the Irish Republic in Dublin. The author of *Wildflowers of the World* and co-editor of *Flowering Plants in Australia*, he sees the role of the Garden as very much that of a living museum, not only of plants but of garden history and architecture. He coordinated construction of a major new tropical conservatory for the Australian Bicentennial celebrations and planned the rehabilitation of a 5.3 hectare site to be added to the Garden when it has been vacated by the State Transport Authority.

The first superintendent of the Botanic Garden was also enthusiastic about garden architecture. George Francis, an English botanist from the prestigious London nursery firm Loddiges, migrated to Adelaide with his family in 1849 and was appointed

superintendent of the new Botanic Garden in 1855. As 'founding father' of the Garden, he designed the first 17 acres (6.9 ha) to incorporate 'a little of Kew, a little of Versailles and a little of what was current in the Dutch and German styles, but not much Australian flavour'. The site was about 43 acres (17.4 ha) of lightly wooded and grassy open land known as the Police Paddock, formerly used for grazing police horses in part of the Governor's domain. It was bounded by North Terrace and the River Torrens, and adjoined the Hospital of Adelaide. With a variety of excellent soils, some swampy areas and some beautiful undulations, it had been chosen by George Francis, together with a committee of the Agricultural and Horticultural Society, as being the most suitable land available. It was still very much as it had been when the first settlers arrived in the colony in 1836, and it was typical of the Adelaide landscape.

At that time the site for a botanic garden had been marked on the city plan made by the surveyor-general, William Light. Its location was on the south bank of the River Torrens and it was obvious before any planting began that the site would be hopelessly flood-prone. The idea was abandoned. In 1837 a second site was chosen on the southern side of the river, extending towards Thebarton in the direction of the gaol (now the railway yards where North Terrace and West Terrace meet). The area remained unfenced and was occupied by Thomas Allen, who advertised 30 000 fresh vegetables and melons for sale during 1839. This site was also abandoned. The third attempt to establish a botanic garden was in 1840, when land was chosen on the terrace below the South Australia Company's brick kiln on the north bank of the River Torrens. Governor Gawler and his wife and a list of ninety subscribers each paid approximately £5 towards the salary of the first colonial botanist for Adelaide, John Bailey, who was to start the Garden. He had been appointed government botanist for New Zealand but because of a delay there was persuaded by Governor Gawler to take the Adelaide position. His salary was £80 per annum.

A management committee was formed and, in return for their subscriptions, the backers were entitled to free admission to the gardens for themselves and their families, and free vine cuttings and seeds. John Bailey enclosed 12 acres (4.9 ha) and established large beds of handsome plants and herbaceous borders. He also planted varieties of fruit and crops he had brought from England, limes, six kinds of olives, blackberries, grape vines and date palms. Within a year the gardens were well established and it was predicted in a newspaper of the day that 'our colonists will enjoy a rich treat in taking a walk through the gardens in our fine summer evenings'. But by the end of 1841 the South Australian treasury under the governorship of George Grey was empty, and John Bailey was retrenched. He started his own private nursery nearby at Hackney and the Gardens land was leased to George McEwin, who grew fruit trees for sale.

There was another attempt, quashed by the colonial architect, to get land between North Terrace and the River Torrens (where the university now stands) for a botanic garden, but it was not until 1855 that the fifth and successful attempt was made. The South Australian economy was for the first time in years comparatively buoyant after a few good farming seasons, and George Francis, who had lobbied the government tenaciously from the time of his arrival in South Australia, was given a grant of £3000 for the year with which to establish and maintain a botanic garden. He was also given a board of eight prominent citizens to administer it, includ-

Far left: Ficus macrophylla, **Moreton Bay fig trees, in Botanic Park, adjoining the Botanic Gardens.**

Left: **A group of** Beaucarnea recurvata **in the Botanic Gardens.**

Right: **Plane tree grove underplanted with a groundcover of hellebores.**

Above: The Australian American Association Memorial on the western bank of Top Lake.

Top right: Nelumbo Pond planted with oriental sacred lotus.

Right: Palms on an island in Top Lake.

ing the Mayor of Adelaide, and later an Act of parliament to protect it. His salary was £150 per annum and he declared that he would 'execute the garden with knowledge, care and zeal so that the Garden shall be as far as my power lies, an honor and benefit to the colony'.

He began by clearing the site of some river red gum, *Eucalyptus camaldulensis*, stumps and filling holes. The land was then enclosed by the government and by the end of three months 52 of the 60 chains (1.2 km) of boundary fence had been erected. The cleared land was trenched to a depth of 18 inches (46 cm) with the help of six men at 7s a day, and planted on the western side with roses of various types. Inside the trenches a belt of Lombardy poplars and Kangaroo Island acacias was planted. By the next year a formal design was being carried out in the Garden. George Francis laid out 'a great circle' in the middle of the Garden, and many of the walks still used today were laid down and trees planted along them. The Upper Lake with its island had been formed from the creek. In September 1857 the Botanic Garden was opened to the general public on weekdays from 9 a.m. to 5 p.m.

Funds were set aside for seeds to be bought from all over the world: £5 went to Germany to Robert Schomburgk, the brother of the future director of the Garden, for specific seeds. Francis began an exchange system of plants and seeds between the botanic gardens of Adelaide and Melbourne, and Hobart and Sydney, and a strong connection was established between George Francis and Ferdinand Mueller, the government botanist at the Melbourne Botanic Gardens. Bulbs arrived from Ghent, seeds from other horticultural centres of Europe, and a request for seeds was made to South Africa after it was heard that Dr Livingstone, the noted explorer, had brought seeds to Capetown.

Before long, bands played in the Garden, visitors promenaded, and the committee set about procuring a £500 conservatory and two greenhouses. An acre and a half (0.6 ha) of the Garden was ceded to the lunatic asylum in the south-eastern corner. Plants were raised in the Garden for Adelaide's parklands and for the grounds of Government House. Francis was also the original landscape architect for all of Adelaide's city squares and for Government House.

By the time he died in office in 1865, Francis had introduced five thousand species of plants into the Garden; these had been catalogued and copies of the catalogue distributed. He had placed statuary throughout, and guns captured at Sebastopol had been positioned in the park opposite the gardens. He had begun the first collection of waterfowl and zoological specimens, which grew to embrace other birds including fifty-two species from India and many animals. Among the birds was the 3 foot (91 cm) high secretary bird, imported from South Africa and noted for its snake-killing ability, but it was unable to breed successfully in the Gardens.

As botanical funds were short, he had had to

provide money for the zoo from his own pocket for some time. The policy of keeping birds and fauna at the Garden continued for the next twenty years. Eventually, after much debate and haggling, the Botanic Gardens Board gave land for a separate Zoological Gardens on the western side of the Botanic Park. George Francis also built a museum in a rustic Greek temple style for displaying botanical specimens, which were later rehoused and added to by his successor, Dr Richard Schomburgk, who took office in September 1865.

Richard Schomburgk was already known as a plantsman. Trained at the University in Freyburg, Germany, he worked in the Royal Gardens and at Sanssouci in Potsdam Germany. In 1837 he accompanied his brother Robert to British Guiana, where they not only saw the *Victoria amazonica* lily, but examined many native South American species including the curare plant used as a poison for Indian arrowheads. For the publication of papers on this, Richard Schomburgk was made an honorary Ph.D. of the German Academy and his future was assured.

In 1849 he and his brother Otto, their wives and families, joined a group of German citizens migrating from Germany in the *Princess Louisa*. They had taken part in political upheavals in 1848 and for that reason had to leave Germany. The *Princess Louisa* sailed to Adelaide, where the Schomburgk brothers became successful vintners at Gawler, making wine from Madeira, Verdeilho and Mataro grapes. To keep alive his interest in science, Richard became curator of the Gawler museum, and from this position he was selected as the second director of the Adelaide Botanic Garden from a field of fourteen applicants.

In his first report he told the Board that he intended to form an experimental garden for the economic use of horticulture and botany; to increase the florist houses; to remove the tiger and bear cages to a more distant part of the garden than their present position; to institute a scientific scheme of planting based on the natural system outlined by Bernard de Jussieu, the renowned French botanist of the eighteenth century and refined by his nephew Antoine de Jussieu in the nineteenth century; to build an aquarium for water plants; and to increase the number of zoological specimens.

Within two years he was able to report the start of the rosery; flower parterres laid out in the fashionable ribbon-style mosaic; the beautifying of islands in the lake; and the imminent building of an aquarium heated by solar power. Already eight species of *Nymphaea*, six of *Nelumbo*, and many other aquatic plants had been acquired from Germany. Two years later he had distributed from the garden fifty cuttings of grape vines and mulberries for silk-growing; and in the Garden, according to the *Adelaide Observer* of 28 November 1866, 'The 40 acres were rapidly being transformed into an emporium of floral beauties of the world'.

By 1868 the Victoria House for growing the *Victoria amazonica* waterlily had been built. The waterlily flowered annually. Schomburgk went ahead with the class ground, planted in the shape of a hippodrome, and by 1874 it represented 130 orders and 750 genera, with every order represented by from four to twenty genera. The greenhouse collection expanded one year by 800 species, and it was obvious that more accommodation was necessary. Using old materials, Dr Schomburgk had built a new stove (a heated glasshouse) to be used for the development of young palms and dracaenas. But there were already about sixty young palms of thirty species for which the present height of the existing stove was inadequate. A larger palm house was necessary.

Richard Schomburgk had heard about a palm house built of iron in Bremen by G. Runge as a private commission for a Herr Rothermunde. It was considered one of the finest structures in Germany. The architect was commissioned to build a similar one for the Adelaide Botanic Garden and it was erected in the Garden during 1875–76. Forty feet (12.2 m) high at the dome, 100 feet (30.5 m) long and 35 feet (10.7 m) wide, with four thousand panes of glass, it was opened in 1877 by Lady Musgrave, the Governor's wife, and deemed 'A fairy palace'. The Palm House held a display of *Latania lontaroides* surrounded by plants of variegated foliage enclosed with gilt and painted tiles in the central rotunda, so that it resembled a 'large and elegant basket of flowers', according to reports of the time. The eastern bay had a fountain and the western bay a display of stalactites from Germany's Black Forest amidst a rockery. This last fernery is still extant.

Richard Schomburgk was also active outside the Botanic Garden. In 1873 the Botanic Garden Board began to plant and fence a tract of land in the former police paddock between the northern boundary of the Garden and the River Torrens. It was to be used as an arboretum, a public pleasure ground and a public meeting-place. It was called Botanic Park. Here Schomburgk planted eight thousand trees, including the Moreton Bay fig trees under which Adelaide's citizens today lunch and picnic. A carriage drive was made through avenues of pines, Moreton Bay figs and plane trees with vistas to St Peter's Anglican Cathedral and the Adelaide Hills. A central area was set aside for bands, concerts, horticultural shows and public meetings. In 1874, the now famous Plane Avenue, the most popular entrance to the Garden, was planted. (Plane trees are among the few deciduous northern hemisphere trees to flourish in Adelaide's hot, dry climate.)

Another major addition to the Garden was the Museum of Economic Botany (opened in 1881), which housed samples of wood grains and resin in the newly designed showcases copied from those in economic botany museums in England. The museum was built in the Attic style with a portico, magnificent natural lighting, and a painted and gilded wooden ceiling. Before long it had a fine collection of botanical objects, with an herbarium at one end. By 1844 Dr Schomburgk was reporting

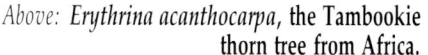

Above: Erythrina acanthocarpa, the Tambookie thorn tree from Africa.

Right: Callistemon speciosus, the Albany bottlebrush, in the native plant section of the Gardens.

Below: The Classground, illustrating plant relationships.

Above left: Bearded iris thrive in the colder climate of the Botanic Gardens at Mount Lofty.

Above centre: Echium and spiraea grow well in the acid soil of Mount Lofty.

Above: Mollis azaleas at Mount Lofty.

Left: The view from the hills of Mount Lofty over the Botanic Gardens.

Right: Conifers and deciduous temperate trees at Mount Lofty.

to the Board that 'The Garden not only forms the centre of civilisation, but it has become a source from which reliable and valuable information has been disseminated'.

In economic botany Schomburgk experimented with vines, mulberries, liquorice plants, sunflowers, castor oil, pyrethrum, kumaras, almonds, hops, tobacco, dried fruits and grasses. He died in 1891 still holding office, having increased the number of species in the Garden from five thousand to approximately fourteen thousand, and laid the framework of the Garden for the next century. Through his work in the Adelaide Botanic Garden he had become internationally famous and was awarded a British knighthood, together with international scientific and political decorations. He had worked closely with his successor, another European-trained botanist, Dr Maurice Holtze from Hanover, who had previously been superintendent of the Botanic Gar-

den in Darwin, which was administered by the South Australian government.

During the last few years of Schomburgk's directorship, development of the Garden had been slow, and a request for an eight-hour day had been persistently refused by the Board. Dr Maurice Holtze's period in the Garden from 1901 to 1917 was marked by social change. Employees were given an eight-hour day and nine days holiday a year, but the prohibition on smoking and wearing buttonholes in the Garden remained. Labourers' wages were increased from 52s to 54s a week. Surplus flowers were given to hospitals and charities. His work in the Northern Territory had mainly concerned experimentation with economic crops and, in addition to a knowledge of tropical aquatic and other tropical plants, he brought to Adelaide the expertise to begin an experimental type orchard. The Board had purchased land at Mylor in the Mount Lofty Ranges in

1898 to grow fruit trees true to type, and Holtze supervised the planting of 5300 fruit trees of 2872 varieties, including 935 different sorts of apples. Stock cuttings from the orchard were distributed to the general public. (The orchard lasted until 1930 when, almost depleted through neglect, it was leased as a cow paddock and in 1941 reverted to crown land.)

In the Botanic Garden Holtze maintained the collection of stove plants and aquatic plants, which remained the best in Australia. He planted the many varieties of wisteria on the east side of what was then the rosery and which now, trained over an arbour, are among the glories of the Garden in spring. Despite the economic crisis of the 1890s, which cut the budget of the Garden, there was a continued wealth of interest in flowers. Wrote Thomas Brooker, the chairman of the Botanic Garden Board in 1909, 'Melbourne and Sydney might beat us in lawns and out-of-doors landscape but Adelaide scored in its wealth of flowers indoor and out, its collection of waterlilies, its cacti and its indoor collection of palms and so on'. By the time Maurice Holtze retired in 1917, the emphasis of the Garden had swung, as it had in botanic gardens throughout Australia, from plant-collecting, botanical science and experimenting with economic crops, to floriculture. The next director, John Frederick Bailey, continued this policy until his term ended in 1932.

His connections with the Adelaide Botanic Garden were intimate, for his grandfather, also John Bailey, had been the first colonial botanist in 1840 and had made the third attempt at founding a botanic garden. His father, F. Manson Bailey, having been trained at his family's nursery, went to Brisbane to become the Queensland government botanist. John Frederick Bailey assisted his father at the Brisbane Botanic Gardens, and was a lecturer at the Brisbane Agricultural College before being appointed both director and government botanist of the Brisbane Botanic Gardens.

In Adelaide Bailey concentrated on flower-raising, and at the height of his period as director some 50 000 floral seedlings were raised annually at the Garden. It was Bailey who built a 61 metre arbour for Holtze's wisteria, the structure still standing today. He redesigned the original Schomburgk rosery as a dahlia garden and grew there annually 1500 dahlia types, 200 of which were given field trials and the best selected for commercial growth. Adelaide's flowers were famous. After a trip abroad in 1921, a member of the Botanic Garden Board reported that, with the exception of the Royal Botanic Gardens at Kew, none of the gardens she had seen possessed collections superior, or even equal, to those of Adelaide.

New land that increased the acreage from 40 to 46 (18.6 ha) was added under the directorship of Harold Greaves, who succeeded John Bailey after working at the Garden from the age of thirteen. An arrangement with the Department of Education resulted in the Garden giving up land on which the glasshouses and the director's residence were situated to the Adelaide Hospital in exchange for hospital buildings – the old Consumptive House and other buildings on the far eastern side of Main Walk. Undulating and higher than the original grant of Garden land, this gave a view over the Garden and a chance for landscaping that the Garden had hitherto lacked. This exchange took place in 1937, and by the 1970s the last of the old hospital buildings had been replaced by the administration block and the State Herbarium.

The herbarium, which in Richard Schomburgk's time had been in the Museum, was revived in 1950 when Noel Lothian, a horticulturist who had worked in Melbourne, Christchurch, New Zealand, Kew in England and the Botanic Garden in Munich, became director of the Botanic Garden. He instituted the State Herbarium of South Australia within the Botanic Garden, one of the most productive herbaria in Australia for both research by botanists into indigenous South Australian flora and for publications.

Under Noel Lothian's directorship, which lasted till 1980, new botanic gardens were added to the existing central one. In 1952 he arranged the purchase of the 44.5 hectares at Mount Lofty for a cold-climate collection of plants impossible to grow on the Adelaide Plain. With a forest cover of stringybark eucalypts, elevation and an acid soil it had a good environment for azaleas, rhododendrons, cold-climate alpines and deciduous trees from both hemispheres. Each year hundreds of new species are planted and already the collections of iris, mollis azaleas and alpine plants are memorable. Its winding roads with convenient viewing spots draw thousands of visitors on weekend drives. Over the years, adjacent land has been progressively purchased and the Gardens area is now 80 hectares, with substantial native regeneration. Planning and development will

Left: Leucadendrons blooming at Wittunga, one of the many South African species grown there.

Bottom left: South Australian and Western Australian sand-plain species displayed at Wittunga.

Below: Leucospermum reflexum.

Right: Long borders of gazanias along pathways leading to the homestead of Wittunga.

take another thirty years to complete according to Dr Morley. Views over the Piccadilly Valley to the Barossa Valley from the tops of the ridges show a countryside as pastoral and green as any English vale.

The second new Botanic Garden, in the Adelaide Hills at Blackwood, is quite different in character, with a dry, windswept terrain. 'Wittunga', of 15.2 hectares, was formerly a farm established in 1901 by grazier Edwin Ashby on land forested with virgin peppermint and blue gum. The Ashby family gave the farm to the state in 1965 when it was already established as a private garden specialising in collections of succulents, and South African proteas and ericas. Edwin Ashby, the son of the founder, continued to live there, to add to the collections and to tend the garden till his death in 1971, when the Board of the Botanic Gardens took over responsibility for it.

In late spring – September–October, the sight of 150 species of South African ericas blooming over undulating paddocks is unforgettable. So too is the collection of seventy-six species of *Protea*, which are also at their height of flowering then. Round the Ashby homestead, the collections of leucadendrons, the specimens of *Leucospermum reflexum* and the *Gazania* borders are in full glory. Over on the furthest hillside of the Garden, informally landscaped round a clump of eucalypts, the blues, yellows and whites of the spring-flowering Australian sandplain natives that have been planted on imported desert sand on top of the indigenous clay soil are spectacular.

Today the Botanic Gardens of Adelaide are under the jurisdiction of a Board of Governors and administered by the Department of Environment and Planning of the state government.

With an estimated 570 000 visitors a year, the Botanic Garden in the city still attracts enormous attention. One of the most exciting projects of any of Australia's botanic gardens, the building of a new tropical conservatory designed by Raffen Maron, Architects, is being built as a Bicentennial project, and it promises to be as notable as the Palm House from Bremen. More than 90 metres long and some 40 metres wide, it will have the capacity to house trees up to 20 metres tall. The design is lens-shaped, and the site, at the Plane Tree Avenue entrance of the Garden, is part of the State Transport Authority Bus Depot.

The spectrum of both exotic and native plants offered by the three botanic gardens of Adelaide was widened with, in 1986, the addition of responsibility for the Black Hill Flora Centre, which was created to enable research on native plant species for the horticultural industry, and relating to endangered species. It is situated in the north-eastern foothills of Adelaide.

With this addition, the botanic gardens of Adelaide fulfil the widest possible purposes for botanic gardens. The aesthetic, passive recreational, educational and scientific roles are balanced. The comment by Anthony Trollope, the English writer who visited the city in 1871, that 'The Botanic Garden of Adelaide pleases like a well-told tale', is still apt.

9

THE DARWIN BOTANIC GARDEN

When Cyclone Tracy devastated the city of Darwin on 24 December 1974, the Darwin Botanic Garden lost 78 per cent of its trees and shrubs. Those that remained were clinging by their roots, their foliage shorn off. The nursery, the fern house and the original curator's cottage (dating from the late 1880s) all disappeared.

The Garden had already survived destructive cyclones in 1897 and 1937, a disastrous fire in 1902, and the bombing of Darwin by the Japanese during World War II. Today it is growing as vigorously as it has ever done. The gaps left by Cyclone Tracy have been filled with new plant collections and, in some sheltered areas, rainforest trees have grown to pre-cyclone heights.

The climate of Darwin helps to account for this speedy rehabilitation. Its monsoonal nature means enormous growth in the wet season from October to May, when the annual rainfall of 1594 mm mainly occurs. The mean maximum July temperature is 28.2 degrees C, with a minimum of 20.2, and the January maximum is 31.9 degrees, with a minimum of 21.8. But the underlying reason for the Garden's survival is that the beauty and strength of the original design, laid down by the first curator of the Garden in 1884, German migrant botanist Dr Maurice Holtze, has been adhered to by subsequent directors and regenerators.

Today Maurice Holtze's blueprint for the 'unbrageous paradise' of the Garden is still paramount: its established atmosphere, deriving from avenues and paths, the rhythmic groups of large shady trees making pools of coolness, and bright, open, sunny savannah grasslands with a few specimen trees. You see it most clearly by entering the Garden, at the bottom of the hill on which it is situated, from Gardens Road, where the main gates are – about 5 kilometres from the centre of Darwin. A variegated hedge of green and white *Acalypha* establishes the formal tone. Overhead is a vaulted canopy of the beauty leaf tree, *Calophyllum inophyllum*, with its thick, shiny foliage intertwined with the feathery foliage of elephants ears, *Enterolobium cyclocarpum*, from the Amazon rainforest, planted on either side of the road. *Pandanus pacificus*, acalyphas and small palms form an understorey with a groundcover of mother-in-law's tongue, *Sansevieria*.

The road approaches a fork, marked by an island of Victorian shrubbery. Fringed with a border of strap-leaved *Hymenocallis*, the white spider lily from South America, and the *Nephrolepis* fern, it is full of palm and shrub planting. The vari-coloured leaves of shrubs are as brilliant as any flowers. Some of the trees, like the travellers palm, *Ravenala madagascarensis* from Madagascar, and the Palmyra group of palms, probably date from Maurice Holtze's day. This island of planting is featured in some of the first photographs of the Garden in the late nineteenth century.

On either side of the road are hidden 'rooms' (enclosures of planting linked by narrow grassy paths) with some old bamboos, *Bambusa vulgaris* and *B. arundinacea*, thickly intertwined; in India they are used to protect villages from tigers. Overhead are large rainforest trees including a fine example of the indigenous lilly-pilly, *Cleistocalyx operculatus*, originally named *Eugenia holtziana* after Maurice Holtze. On the left side of the road, towards the open savannah, shrub and tree plantings include a mature *Beaucarnea recurvata*. In the distance are groups of *Ficus benjamina* and the rain tree, *Samanea saman*, from the Amazon. Past the fork in the path there is another pool of shade from large *Samanea saman* trees, and the road goes up the hill past post-cyclone collections of casuarinas, crotons and *Latania* palms to the administration centre at the top of the hill. From here you look out over the thick tropical greenery of the Garden to the milky blue of the Arafura Sea off Mindil Beach.

The Gardens were originally 80 acres (32.4 ha) when they were started by Maurice Holtze, but more land than was needed was gazetted for a future botanic gardens and parkland. Since Cyclone Tracy, examples of rainforest, wetland, sand-dune and lagoon ecosystems have been added to the Gardens, making them approximately 98.8 acres (40 ha). With the exception of the parklands of the Cairns Botanic Gardens, this is the largest botanic garden in the

---

Left: **Avenue of royal palms,** *Roystonea regia.*

north. Only 11 degrees latitude from the equator, Darwin also has Australia's most northerly botanic garden, and the only one with a monsoonal climate.

The land for the Garden, originally known as Paper Bark Swamp, was selected as being the most fertile around Darwin. Of all the colonial outposts, Port Darwin, or Palmerston as it was called in 1869 when it was founded, was most in need of horticultural and botanical expertise to grow food and viable crops for export in order for the colony to survive. The climate has no temperate features, and with virtually no rain for six months of the year, it is difficult even today to grow vegetables all the year round. Eighty per cent of the Northern Territory lies within the tropics. The hinterland is composed of rocky outcrops and mangrove flats threaded with tidal rivers. Other land has poor or variable soil. The first settlement in the Top End, at Port Essington in 1824, was abandoned in 1849 when thirteen of the fifty-two people there died and another fourteen were ill.

When Darwin was established on 5 February 1869 by the surveyor-general G. W. Goyer, on the order of the premier of South Australia, the Government Garden was laid out within a week. Fred Schultze, the first gardener, had brought seeds of tomatoes, cucumber, kidney beans and pumpkin for a population that by the next year was fifty. The Garden was at Frances Bay, and with William B. Hayes as head gardener, crops of tapioca, coconut, bananas and, from the Botanic Garden in Brisbane, sugar cane, were grown. By 1876 indigo and fibre plants (probably sisal) were added to the list, and 9 acres (3.6 ha) of sweet corn were also cultivated.

With food and crops established, the need arose for a public pleasure park with trees and flowers. William Hayes died and Maurice Holtze, who had migrated to Darwin with his family in 1872 and taken a job as a warder in the Fannie Bay gaol, applied for the position of 'a fair botanist and florist' advertised by the resident-general Edward Price in 1878. Holtze won the position over Angus McDonald, a candidate chosen by Dr Richard Schomburgk, the director of the Adelaide Botanic Garden, because he could supply his own house and McDonald, coming from Adelaide, could not. They both asked for a salary of £200 per year, and it was approved. Later Holtze asked for a salary increase to £400, which was refused by the authorities.

Dr Maurice Holtze's credentials were impressive. He had done his botanical and horticultural training in St Petersburg, Russia, with Professor Lennis, a famous botanist of the day, at what is now the Kamarov Institute, one of the leading Soviet horticultural centres. Later he had worked in the Royal Gardens at Herrenhausen, near his native Hanover, as first assistant gardener, and for four years at the Hildesheim Gardens nearby in Hanover. He had also worked in a nursery near Hanover for four years.

On his appointment as first curator of the Garden, Maurice Holtze changed the site from Frances Bay to 30 acres (12.1 ha) of land at Fannie Bay. The soil was not outstandingly fertile, but it was fairly representative of farming country around the town and was thought to be the most suitable for plant-testing. In March 1879, 350 Chinese labourers were paid 1s a day to clear the land of dense tropical rainforest and to sink seven wells. By September that year all 30 acres were planted with tropical fruit trees, arrowroot and bananas. While sugar cane was thought to be unviable due to the poor soil, Maurice Holtze recommended adding manure to the soil.

Although he was officially curator of the Botanic Garden, Dr Holtze was primarily occupied with crop-testing. He grew *Sesamum indicum* and extracted oil from the seeds with his wife's mangle. Entries in the Sydney Exhibition of 1880 produced a gold medal for a cotton crop. At the Calcutta Exhibition of 1884 the Darwin Botanic Garden's entries of tapioca, arrowroot and bananas won gold medals. But under constant cropping the soil was becoming exhausted, and in 1883–84 the garden at Fannie Bay was abandoned and the land sold for building blocks.

The new 80 acre (32.4 ha) site nearer to town

KEY TO MAP

1 Future Rainforest Development
2 Pond
3 Rainforest Development Area
4 Ponds and Waterfall
5 Nursery
6 Established Gardens
7 Amphitheatre
8 Plant Display House (Orchids)
9 Residence
10 Holtze Cottage
11 Coastal Sand Dunes with Salt Tolerant Plants
12 Arnhem Land Natives and Savannah

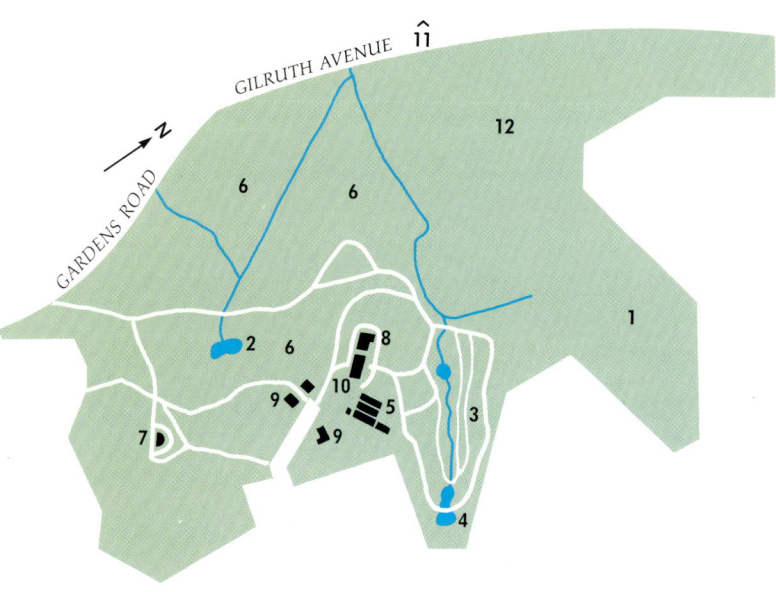

had the advantage of being close to the prison, and thus prison labour, for the founding and maintaining of the Botanic Garden. It was often a dubious advantage, for Maurice Holtze was frequently to complain about the quality of the work. The land had been used for vegetables and crop-growing by Chinese squatters who had no official claim to it. On 1 October 1896, an official proclamation for a botanical garden was issued in order to remove the remaining Chinese squatter, and 10 acres (4 ha) of 'Paper Bark Swamp' were cleared and planted. Maurice Holtze and his staff of seventeen Chinese labourers moved 237 trees from the old site at Fannie Bay to the new gardens, losing only fourteen of them in the transplanting, and by the next year a report of the Garden's progress noted that it was 'fast assuming all features of a well-loved public garden'. Maurice Holtze was growing ornamental trees in vast quantities, and anyone who wanted them was given some. The greenness of Darwin's suburbs today is attributed to the Darwin City Council's continuing policy of 'free trees'.

The cyclone-surviving trees of the Garden today reflect Maurice Holtze's botanical and design flair. Imported from the tropical zones of the world, with a few local species planted, they usually have some economic importance. In the lower part of the Garden, your eye lights on a large Burmese rosewood, *Pterocarpus indicus*, a valuable timber tree. With yellow panicles of flowers and interesting dusty brown seed pods that lie on top of the foliage, it is also useful as a decorative tree. There are, close by, large specimens of the Indian mast tree, *Polyalthia nitidissima*, and the crepe myrtle, *Lagerstroemia indica*, both ornamental in form and blossom. There is a large *Bauhinia blakeana*, with port wine flowers, and an old specimen of lignum vitae, the tree with the world's heaviest wood, and blue flowers during the dry season. Also decorative is the diva diva tree, *Caesalpinia coriaria*, from Patagonia. An *Emblica indica* from India is not only attractive but also known as a valuable source of vitamin C. Several specimens of the golden shower tree, *Cassia fistula*, are not only decorative but are capable of yielding tannin, soap and timber.

For their ornamental quality, Colville's glory,

Colvillea racemosa, with its blossoms of scarlets and the curious *Kigelia pinnata*, with metre-long seed pods resembling sausages, are all well chosen. Poincianas, *Delonix regia*, are also thickly planted in the old part of the Garden. Standing in isolation in an open lawn is a very old specimen of the Western Australian bottle tree, *Adansonia gregorii*, one of the nine species of *Adansonia* planted in the Gardens.

By 1886 some four hundred different crop plants were being tried, including eight sugar cane varieties, and some rubber species in 16 acres (6.5 ha) of planting. At the Calcutta Exhibition of 1887, the Darwin Garden won prizes for indigo, cotton, jute, ginger and tapioca. So important were the Eastern crops to the potential economy that in February 1887 the government resident, John Langdon Parsons, and Dr Holtze journeyed to Canton, Hong Kong, Macao, Saigon, Singapore and Batavia to learn more about tropical agriculture. Arriving back, Maurice Holtze tried crops of peanuts, three crops of tobacco and more cotton. In Saigon the rice-growing had impressed him and he thought rice would be a viable crop for the Territory, but no progress was made with this idea. 'I have vigorously and repeatedly recommended that the Territory grows rice', he wrote. The opium poppy was tried but it was not successful.

By the next year Maurice Holtze had applied

for a grant of £800 to develop the Garden and crops, but due to a financial crisis funds were cut back. He was appointed official forester for the Northern Territory in 1888 and tried to start a scheme to grow the native cypress-pine, *Callitris*, which is resistant to white ant attack. He tried exotic grasses, including Natal grass and Guinea grass.

Holtze did not neglect to explore the hinterland, looking for native species and noting exotic species that had become naturalised. On a trip to Melville Island and Herman Island he listed exotic weeds from gardens established when Matthew Flinders and Philip King had made visits there in 1803 and 1818 respectively, and in 1889 he was made a fellow of the Linnaean Society for his services to botany.

When he took over the Government Gardens their use had been principally to provide fruit and vegetables for the officials of the colony. But, he said, 'The raising of cabbage heads is not the true function of botany'.

However, when Holtze resigned in 1891 to take over the directorship of the larger and more prestigious Adelaide Botanic Garden after the death of Dr Richard Schomburgk, none of the crops that he had assiduously tried had taken off as an answer to the economic problems of the Territory, for the climate and the conditions were not yet sufficiently understood.

Holtze's son Nicholas, who had been born in Russia but educated in Adelaide, and started his

Left: The largest species of bamboo in the Garden, *Dendrocalamus giganteus*, is part of the old planting. This species is used for making furniture.

Top right: This boab tree, *Adansonia gregorii*, survived Cyclone Tracy.

Right: The poinciana tree, *Delonix regia*, one of the many superb flowering trees in the Garden.

working life as a messenger with the Overland Telegraph Office in Darwin, succeeded him as curator of the Darwin Botanic Garden. The salary was £50 a year, the job was that of part-time curator since the economy could not support a full-time specialist, and Nicholas Holtze was only twenty-three years old. But he did seem to know about botany and horticulture. He planted an avenue of coconut palms, *Cocas nucifera,* half a mile long at the bottom of the Garden. 'In a few years it will be a splendid sight', he wrote. He continued to plant large trees from other tropical areas of the world and to nurture the gardens as well as to continue his father's experiments with crops. So difficult was the financial situation that he hoped the Gardens would become self-supporting through the growing of crops, and at one stage he grew thirty thousand sisal plants for distribution to local farmers. But the sisal proved to be unstable in the climate, as did cotton, and he concentrated on rice-testing, thinking that the variety 'Upland Grade I' from America would be suitable for cultivation in Darwin. Because of the lack of reliable labour, a machine-harvestable rice would be the most practical variety.

In 1911 Holtze set out to test this rice crop, but the one person capable of working the machinery, a Mr Milton, broke his arm attempting to start it and the crop had to be harvested by sickle. Nicholas Holtze was so keen to promote rice-growing that at one time there were thirteen varieties in the Garden for testing. As his botanical work was only part-time, he was promoted up the civil administration ladder, becoming the government secretary for Darwin and later acting resident. In 1913, at forty-five years of age, he died. The Holtzes, father and son, had urged important innovations for the economy of the Northern Territory including longer trials for crops, the importation of cheap labour to harvest and maintain them, and government assistance to farmers. None of these things happened, however.

The Botanic Garden Nicholas Holtze left is said to have been beautiful, but only Maurice and Nicholas Holtze knew the names of the plants, and labels were few. A trained botanist and curator was needed to replace Nicholas Holtze and when the position was advertised, C. E. F. Allen, who had been trained at Kew Gardens and had worked for ten years in South Africa, won it. A year later he joined the Australian Expeditionary Force and departed for World War I. Returning at the end of the war he reported that 'These beautiful gardens have become rank wilderness'. He cut out dead trees, made borders and restored lawns. Crops were still grown for experimental development, but there was so little civic money available for the Garden that only maintenance could be afforded. Allen resigned in 1935 and a couple of years later a botanist from the Sydney Botanic Gardens, K. C. Mair, was appointed as curator. (He subsequently returned there as director.)

But with the cyclone of 1937 and in 1939 the start of World War II, which led to the stationing of soldiers in the Botanic Garden area, the Garden deteriorated rapidly. It is reported that six calves lived in the shadehouse for a time, and the whole area was covered with barbed wire. The officer in charge of Darwin's administration, Brigadier E.M. Dollery, M.C., was shocked at the state of the Garden and towards the end of the war began to rehabilitate it. He wrote personally to army headquarters in Queensland, Victoria, Western Australia and South Australia for suitable shrubs and trees for Darwin. When civil administraion was restored in 1945, the Garden was in a more respectable condition.

With the building of a sound shell and amphitheatre on land in the southern part of the Garden, it became the main venue for any large theatrical, musical or municipal event, including Royal visits. But as a botanical institution it languished through lack of funds. In 1957 an experimental farm was begun at Berrimah, near Darwin, and an herbarium, now located at Palmerston, was opened. Today it holds twenty-seven thousand Australian specimens (mainly NT plants) and eighty thousand specimens of exotic plants. These figures include specimens in a Northern Territory herbarium at Darwin, which collects plants found mainly in the Darwin Gulf and the Barkly Tableland area. There are about four thousand indigenous species

known in the Northern Territory area, but it is estimated that there are numerous species yet to be found. Two botanists are employed full-time in scientific research.

The main Garden suffered through lack of funds, although an attempt was made to keep up scientifically with other major gardens in Australia and the nursery and propagation programmes were active. After Cyclone Tracy, when a count was done of irreplaceable lost species, there was only one that could not be re-established from either seeds collected at the Botanic Garden or young plants propagated in the nursery: the tree *Pachira aquatica*, the 'devil's tooth', was missing. Of the estimated 1344 species grown in the Garden, 214 were indigenous plants and 1130 were exotic. The palm collection was surprisingly poor, with only seven species.

With extra money that was allocated for the Garden's rehabilitation after Cyclone Tracy, the present director for Parks and Gardens, George Brown, who administers the Darwin Botanic Garden, has begun a taxonomic planting programme. The present palm collection is of 280 species, making

it one of the most comprehensive to be seen in an Australian botanic garden, even though it is yet in its infancy. There is, on the slope down to the bottom of the gardens, a collection of thirty-five frangipani species. A collection of tropical climbers and creepers is planned for Aboriginal totem poles. For the first time *Ficus benjamina*, a popular and lovely indoor pot-plant, and a native of the Northern Territory, is grown in the Botanic Garden as a decorative feature tree.

Where Maurice Holtze's original cottage once stood, a new shadehouse for orchids was built after the cyclone of 1974. This now houses a collection of orchids donated by John Womersley, the curator of the Lae Botanical Gardens, New Guinea. There is a collection of native Northern Territory orchids, including *Vanda* and *Dendrobium*. In the open area stretching down to the beach the post-Tracy planting scheme will add a recreational facility to the Mindil Beach area. It has been funded and organised by the Northern Territory Government through the Conservation Commission and by the Darwin City Council, which administers the present Garden area

Top: Dendrobium affine, a Northern Territory native orchid.

Above: Phalaenopsis hybrid orchid in the John Womersley orchid section of the Garden.

Right: Nymphoides indica, the flower is dwarfed by its large, green leaves.

from ratepayers' funds. The 3.4 hectare rainforest gully, which has already been planted with 140 rainforest species, features fast-growing leafy cover placed to protect more vulnerable species, and a pool, waterfall and stream falling down the hillside; the lagoon area of 9.6 hectares, for wetland plants to encourage water birds, together with the hillside area, will be planted with Arnhem Land natives and savannah species; the coastal area of 6 hectares has a tidal creek and dune area for the experimental planting of salt-resistant water-loving plants. For the first time the Garden has a planting policy based on ecological and genus collections.

Yet despite all the new areas and new planting, George Brown's design policy is still similar to Maurice Holtze's founding philosophy: that cool islands of dense planting for shade, interspersed with open grassy places of mown tropical paspalum, cope best with the climate.

Ideally the Botanic Garden should become a

territorial responsibility, with facilities to enrich it scientifically. Labelling is still inadequate, and there has been no real census of species in the Garden. A start has been made on uncovering the details and planting of the original Maurice Holtze garden, however. In 1986, for the first time, a circular rockery featuring stones embedded with sea shells was uncovered; it is thought to date from the founding of the garden. Edges and drainage pipes and paths are constantly being uncovered. Because of its geographical position as the only Australian botanic garden with an established collection of plants in the monsoonal tropics, the Garden is of great importance, according to the Royal Australian Institute of Parks and Recreation. The fact that it has the potential to increase the collection of rainforest, tidal and lagoon plants adds to its significance. And the Garden also has the advantage of a lyrical design laid down by Maurice Holtze and respected still in the 1980s.

10

KINGS PARK & BOTANIC GARDEN, PERTH

For the brilliance of its spring flowers, the Botanic Garden in Kings Park, Perth, high on the bluff of Mount Eliza overlooking the junction of the Swan and Canning Rivers and the city skyscrapers, is supreme among Australian botanic gardens.

From August to December, the flowering of three thousand species of plants from the world's Mediterranean climate zones, unfolds on the edge of the 400 hectare park, two-thirds of which is natural bushland and some of which is devoted to a ceremonial park and open parkland. The Botanic Garden, set in grassy paths and planted mainly according to geography, occupies 17 hectares of it. From California there are the varying blues of many species of *Ceanothus*, with the orange of the clumps of Californian poppy, *Eschscholzia californica*; the Mediterranean countries produce the rosemaries, the cytisus, the lavenders and the roses; from South Africa there are the Cape bulbs, the leucadendrons and proteas, the gazanias, gerbaras and pelargoniums; from Chile there are the spectacular blue spikes of *Lobelia urens* var. *brevibractea* and species of *Sisyrinchium*, an iris-like flower; from Australia, mainly Western Australia but including plants from South Australia, are about 1500 species of predominantly ornamental flowering plants. All the species thrive in the Perth climate, which is typically Mediterranean. A mild, rainy winter, with an average annual rainfall of 686 mm and a July average maximum temperature of 17.6 degrees C and minimum of 9.2 degrees, is followed by a dry summer with a January average maximum temperature of 30.3 degrees, dropping to a minimum of 9.2 degrees.

The best time to see the Australian plants is during 'Wildflower Week' in the Garden, which is usually the last week in September or in early October when sheets of the everlasting daisy, *Helipterum roseum*, spread in an impressionist mist of pink, and you will find *Eucalyptus rhodantha*, with its broad silver leaves and lovely pink-red blossom, and *E. caesia*, a weeping, silver-leaved, pink-flowering small gum, together with hakeas, dryandras, banksias, acacias, grevilleas and callistemons in forms rarely seen outside the state.

Although spring is the best time for the Western Australian natives, there is some flowering of bank-
sias, dryandras and grevilleas throughout the year, and with irrigation the natural spring flowering time for most species has been extended into late spring and summer. December–January bring the brilliance of the red flowering gum, *Eucalyptus ficifolia*, and in February there is a second burst of flowers. The copper cups shrub, *Pileanthus peduncularis*, is covered with coral bloom at this time; liberally planted through the Western Australian section, this is a joy to see.

In specialising exclusively in geographically related planting, Kings Park and Botanic Garden is unique among Australia's botanic gardens. Its primary role, though, is to display and study the flora of Western Australia, which to botanists is among the world's most exciting. The state occupies a third of the continent and supplies about eight thousand of the estimated twenty thousand Australian species, plants that are more vivid and bizarre in form than those found in other parts of Australia because of the astringent, changeable climate deriving from isolation by deserts and the seaboard. Here plants and bird life evolved together, and one-fifth of the flora (including *Anigozanthos*, the kangaroo paw) is bird-pollinated.

In 1953 an American Fulbright Fellow, Dr W. S. Stewart, who was studying at the University of Western Australia on leave from his position as director of the Los Angeles Botanic Gardens and Arboretum, pointed out that Western Australia's most urgent need was for a botanic garden where its unique wildflowers could be preserved and studied. With the loss of indigenous flora due to land-clearing for mining and farming, and the consequent increasing salinity of the soil, many Western Australian species were in danger of extinction. The famous Western Australian flowering gum, *Eucalyptus ficifolia*, is now confined to the Walpole area; *Eucalyptus caesia*, one of the loveliest of the small native trees, is restricted in the wild; the black kangaroo paw, *Macropidia fuliginosa*, is becoming rare in its natural habitat. Trial plantings of natives

Overleaf: **A swathe of spring-flowering *Helipterum roseum* in the Kings Park and Botanic Garden on Mount Eliza.**

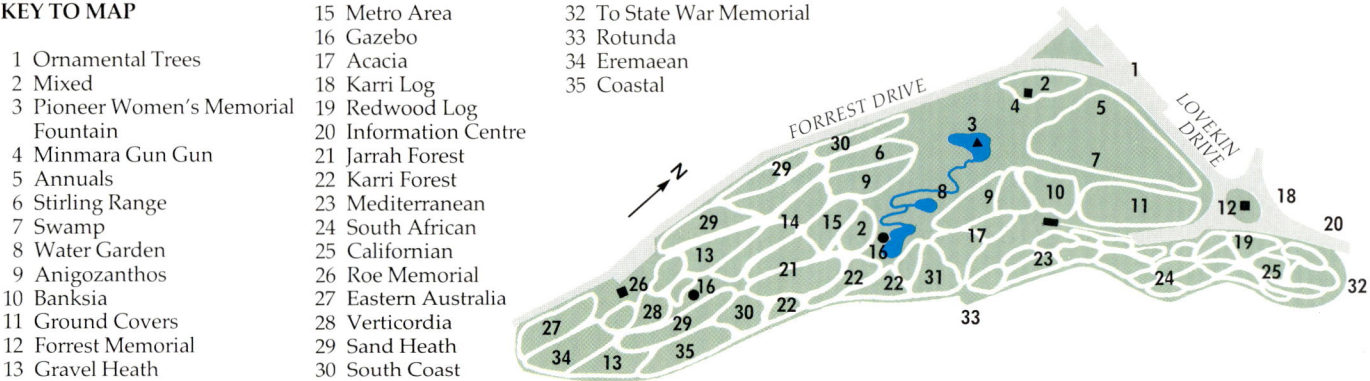

had been made in Kings Park in the mid-1950s, in response to an investigation into the possibility of a botanic garden by the State Gardens Board. It was Dr Stewart's report, however, that convinced the government of its necessity.

On 1 December 1959, a plan for a botanic garden in Kings Park was approved by the government of Western Australia, and in 1961 the first director, Dr John Beard, was appointed. A botanist who had previously worked in America and South Africa, he had studied the Botanic Garden of South Africa at Kirstenbosch, Cape Province, where the flora of the Cape Province and of South Africa is grown almost exclusively, and proposed to grow in Perth not only the indigenous flora of Western Australia but also that from associated climatic regions of the world, so that similarity of form could be observed.

Kings Park, less than 2 kilometres from the centre of the city, was the obvious choice for the site of the proposed gardens, and work began in 1962. The design was a collaboration between John Beard, the director, Arthur Fairhall, the superintendent of the park (who had worked at the Kirstenbosch Botanic Garden), and Professor Brian Grieve, who held the chair of botany at the University of Western Australia. The planting was concentrated along the scarp of Mount Eliza, a spectacular bluff rising above the flat land on which the city stands. Planting down the cliff would make the Gardens inconspicuous from the top of the park and would not obscure the views of the Swan and Canning Rivers and the city below.

Though the exotic plant species are arranged in accordance with their geographic origins, the Australian plants are mostly arranged according to ecological association rather than botanical order, so that there is a mixture of growth forms and taxonomic associations as occurs in nature; thus there are beds for plants from the sandy heath area; the gravel heath; the Darling Ranges; the coastal area etc. The banksias are all grouped together, however, and so are the acacias.

Two-thirds of the Australian plants are from Western Australia, and they occupy 14 of the 17 hectares. Most of them are from the south-western corner of the state, an area extraordinarily rich in species, two-thirds of which are endemic in the re-

gion. They are the plants most threatened with extinction, for development related to mining and farming is proceeding rapidly here, and the saline content of the soil is rising through clearing of the land. This is also an area that is more easily accessible to botanists and horticulturists from the Garden than the desert and far-northern areas are.

The climate of the south-west region is largely compatible with that of Perth and some of it has soils similar to the strongly drained sandy soil in Kings Park. This is up to 30 metres deep, and along the scarp area there are limestone conglomerate outcrops. Since moisture retention of the Kings Park soil is slight, many famous Western Australian plants, like the york gum and salmon gum from the city soils of the wheat belt, cannot be grown easily. Even the famous *Lechenaultia biloba* can only be grown in the park under simulated laterite conditions.

The species that can grow well are introduced into the natural vegetation of the Botanic Garden: amongst the overtopping indigenous jarrah, *Eucalyptus marginata*; the marri, *E. calophylla*; the tuart, *E. gomphocephala*; the medium-sized *Acacia saligna* and *Allocasuarina fraserana*; the low-growing cycad *Macrozamia reidlei*; the black boy, *Xanthorrhoea preissii*; and the flowering shrubs that have become naturalised in the Botanic Garden. In the spring, over the escarpment, you can see the white, yellow and pink of *Dryandra sessilis*; *Kunzea baxteri* and *Chamelaucium uncinatum*, the Geraldton wax, and throughout the beds a meandering of naturalised kangaroo paw, *Anigozanthos*; helichrysums and hardenbergias give a relaxed bushland ambience to the Garden. There is a consciousness of prolific birdlife, with approximately sixty-five native species throughout the park.

Many of the plant species that cannot be grown in the natural soil of the Botanic Gardens are cultivated in four display glasshouses, which were opened to the public in 1985 after five years of planning and building. These are the achievement of the present director, Dr Paul Wycherley, a botanist origi-

Right: **Hibbertia hypericoides**, the native buttercup, grows wild in the Botanic Garden and is here alongside a *Xanthorrhoea preissii*, a black boy plant.

Top left: *Petrophile linearis.*

Top right: **Banksia ashbyi.**

Bottom left: *Dryandra sessilis.*

Bottom right: **Eucalyptus macrocarpa.**

nally from England who has worked in Holland and Malaya and who succeeded Dr John Beard in 1971. Arranged in the shape of a square, the glasshouses form a complex around a courtyard and pool, with landscaping around the perimeter. Each glasshouse has a floor area of 148 square metres, and artificial environmental conditions are simulated to suit the type of planting.

There is a Dry Inland House containing plants from the Eremaean Provinces, which includes the Pilbara, the Nullarbor, and the Sandy and Gibson Deserts. Many of these include saltbushes and other salt-tolerant plants as well as the small annual flowering plants which, after rain, produce a floral carpet in the desert. The Carnivorous Plant House contains not only insect-eating species from Western Australia, for which the state is famous, but also similar plants from similar climates elsewhere in the world, including the Venus fly trap from North and South Carolina and numerous American pitcher plants. This is combined with a collection of halophytes and Pilbara region collections of plants, which are species growing in saltpan areas and highly salt-tolerant. They include numerous saltbushes, bluebushes and samphire species.

The Kimberley House, accommodating plants from the arid-monsoonal climate of the Kimberley region in the north of the state, is interesting to visitors as it displays flora quite different from that of the rest of Western Australia. In type it is linked with the flora of northern Australia, and the Kimberley region grows some 1500 indigenous species, of which 1200 do not grow to the south of it. These include the famous boab trees. The glasshouse contains many smaller plants of the area – *Hoya australis*, an attractive scrambler, and the lily *Crinum flaccidum* among them. The Fern House, with a collection of ferns from throughout Australia, includes the primitive fern *Marsilea*, with four-lobed leaves similar to those of a four-leaf clover.

Brick paths link the glasshouses. At the back of them the paths are grassed. In the central courtyard the pool grows aquatic and swamp plants and the pergola is hung with Australian climbers: *Hardenbergia comptoniana* and *Kennedia macrophylla*.

Of the planted beds around the glasshouses, those containing the lechenaultias, red and yellow and the famous blue *Lechenaultia biloba*, are the most floriferous, with the *biloba* species flowering in the spring, the other colours throughout the year. A South African bed includes a curious plant, *Welwitschia mirabilis* from the Namib desert, a succulent

Top left: *Anigozanthos manglesii*, the emblem of Western Australia.

Top right: *Macropidia fuliginosa*, the black kangaroo paw.

Bottom left: *Diuris longifolia*.

Bottom right: *Eremaea fimbriata*.

with corrugated leaves; there are other succulent desert plants too. There are beds for species of cactus and of euphorbia, an eremophila bed and a verticordia bed, which includes *Verticordia grandis*, with large red flowers, from the northern sandplain.

Another new development in the Garden is the establishment of arboreta containing species indigenous to the area. These are valuable, as most of the bushland in the park consists of secondary growth owing to frequent fires and the culling of timber by the people of Perth until the 1860s.

The decision to break into the bushland of Kings Park (called Eliza Park until 1901) and to construct the Botanic Garden roused some opposition. Ever since the park was first gazetted as a recreation space for the public in 1872, it has been treated as sacrosanct. It was one of the first great areas of natural bushland gazetted for public recreation anywhere, and although given token fencing and closed at night to vehicles up until 1940, it has never been entirely closed to the public and is now unfenced. The problems of having a botanic garden open to the public twenty-four hours a day are worrying, but over the years the objections have proved groundless, although there is a high loss of plant labels and some loss of plants.

A second accession of land in 1890, to mark the inauguration of responsible government for Western Australia, was gazetted by the first premier, John Forrest, who had worked as an assistant surveyor with the original surveyor-general for Perth, Lieutenant John Septimus Roe. This addition brought the park to 397 hectares.

Since its inception the Park had, for the people of Perth, taken the place of the traditional botanic gardens and pleasure parks of the eastern states, like the rings of greenery round Sydney, Hobart, Brisbane, Melbourne and Adelaide in their picturesque style. Premier Sir John Forrest, although he is reputed to have admired the Western Australian flora, had more fondness for the oaks and elms and an English-style landscape, and in 1892 he and the managing board managed to get £2000 granted by parliament for improvements to the Park for carriageways and plantings. Oaks grown from acorns brought from Windsor Great Park in England failed to thrive, as did many other northern hemisphere trees, and lawns, with the poor water retention of the Kings Park soil, were futile, but avenues of pines, palms and *Eucalyptus ficifolia* became well-established and many of the trees still thrive today. Under Premier Forrest's guidance, a further grant by par-

The parkland of Kings Park.

liament of £17 000 was made for the buying back of land along the scarp – today's Botanic Garden site – which had become alienated from the rest of the parkland. In 1897 the Terrace Gardens, pleasure gardens on the scarp cliffside going down to Mounts Bay, was built.

With a rockery, terraces, a 'Fairy Grotto' and 'Lovers Walk', plantings of exotic trees and flower beds and a walk over a bridge and down terraces to a tea house by Mounts Bay Road, it had some of

Above: **The silver foliage of** *Eucalyptus rhodantha.*

Right: **The bushland of Kings Park, mainly secondary growth, is one of the world's earliest natural public parklands.**

the characteristics of a Victorian pleasure park. Trees were mainly from countries with Mediterranean climates: *Pinus halepensis*; the cork oak, *Quercus suber*, and jacarandas. And there were agaves, cacti and *Dracaena draco*, popular in the Victorian era.

The grotesque appearance of the limestone rock face distressed some members of the park's board, and the director of the Melbourne Botanic Gardens, William Guilfoyle, was asked his opinion on beautifying them when he visited Perth in 1907. He recommended cacti and succulents for the rock face. He was astonished and delighted at the indigenous quality of the bushland in the park and wanted seeds for his Melbourne Gardens. He urged that the indigenous flora be encouraged and increased, but exotics continued to be grown until 1919.

The visit of the director of Melbourne's Botanic Gardens did not bring about the establishment of the true botanic gardens in Perth, however. One possible reason is that the population of Perth was slower-growing than that of other states; by 1863 it was barely 15 000. Perth was also the poorest of the mainland major cities, and only after the 1890s gold rush was there money for a museum, a treasury, an art gallery and zoological gardens.

Rare trees and exotic flora were grown, though, in the Treasury Gardens, and in the gardens of Government House, which had been planted by Surveyor-General Roe, a keen seed-collector and plantsman. And in the city there were formal parks like Queens Park, with specimen trees and duckponds.

In the Acclimatisation Gardens for Zoological Species there was a good collection of exotic plants including palms. But there were no official botanic gardens, although there had been attempts previous to that of Sir John Forrest to start them.

Accompanying the first 150 settlers aboard the *Parmelia*, who arrived at the Swan River Colony in 1829 under the direction of Captain Stirling, the governor of the new colony, were all the necessities of life: cattle, sheep, pigs, poultry, and nine boxes of plants including apples, cherries, plums, pears, peaches, vines, gooseberries, currants, strawberries, Jerusalem artichokes, dahlias and chrysanthemums.

Aboard the *Parmelia* was James Drummond, the honorary government naturalist of the new colony, who had set sail on the understanding that, if a public garden were started, he would be the salaried superintendent. A trained botanist and horticulturist, he had, at the age of twenty-two in 1804, been curator of the Cork Botanic Gardens in Ireland. When funds for these were withdrawn by the British Government he was pronounced redundant, and applied to go with Captain Stirling to the Swan River Colony with his wife and six children.

On arrival in Western Australia, in his capacity of honorary naturalist, he planted the seeds and set out the plants at Garden Island and, with other colonists, was urged to select 5000 acres (about 2000 ha) of land for £375. Leaving the new garden established and in the charge of a gardener called Morgan, he applied for a grant of land at Guildford (at the junction of the Swan and Helena Rivers), where he planned to establish a public garden. In 1829 he asked Governor Stirling's permission to transfer some of the stock from Garden Island, but the request was refused. From Guildford he moved twice in the attempt to find a site for a government garden, the second time establishing his own nursery on 2 acres (0.8 ha) at the foot of Mount Eliza.

This decision probably led Governor Stirling to start a public garden adjacent to Government House and to appoint Drummond as superintendent of it with a salary of £100 per year. Drummond worked keenly to set out a garden, and it was recorded by George Moore of the Agricultural Society on 3 July 1831 that 'A Botanical Garden has lately been laid out, in which I walked with the Governor and his Lady accompanied by some of my kind friends'. But by 1832 funds for Drummond's salary were withdrawn, and Governor Stirling suggested that Drummond take over the new gardens for his own profit until the matter was further sorted out. Drummond agreed, on the condition that he had a residence that could double as a storage place for his seeds, plants etc., near the gardens. The governor complied with the request.

Governor Stirling having gone to England to discuss the problems of the new colony in 1832, James Drummond was left in charge of the gardens for the next two years, maintaining them as best he could from the sale of their produce.

On the Governor's return, Drummond was offered the lease of the gardens. He accepted it and the role of superintendent, but requested that his house, situated under a tree in the gardens, be included in the lease. Governor Stirling, who had plans to move the existing government house to the site of the gardens, asked Drummond to relinquish the house. Drummond suddenly lost heart and wrote his resignation from the Government Gardens. He retired with his family to a 2900 acre (1173.5 ha) grant of land in the Helena Valley to grow olives and vines.

But in 1841, with the appointment of a governor John Hutt, James Drummond again presented his case for Government Gardens. The old gardens near Government House had not survived long after Drummond had withdrawn.

Governor Hutt agreed that the project was desirable and offered Drummond a piece of land near the old Government Gardens, for which Drummond was to bear the cost of clearing. Drummond became

Left: Nineteenth century wall in the Terrace Gardens within the Park.

Right: The Limestone grotto within the Terrace Gardens built in 1897.

Above left: A bed of *Conostylis setigera*, a relative of the kangaroo paw.

Left: Hardenbergia entwined with *Hibbertia hypericoides*.

Above: *Ceanothus thrysiflorus*, the Californian lilac.

Top right: *Eschscholzia californica*, the Californian poppy.

Middle right: *Isopogon dubius*, the 'pincushion coneflower'.

Bottom right: *Chamelaucium uncinatum*, the Geraldton wax plant.

involved in planning botanical expeditions into the wild and did nothing about the clearing of the land: concluding that Drummond had changed his mind about the project, Governor Hutt did not get in touch with Drummond again. Drummond continued to work as a botanist, collecting seeds and plants from the wild for export. He collected for the Royal Botanic Gardens, Kew, and the Royal Horticultural Society, as well as commercial nurseries and private garden collections. Though he failed to found a botanic garden for the colony, he did help to establish a scientific base in England and Europe for Western Australian flora.

Although by 1904 there was still no official botanic garden, it was considered necessary to establish an herbarium to concentrate on the scientific aspects of the local flora. Today the Western Australian Herbarium is administered by the Western Australian Department of Agriculture and contains over 300 000 dried specimens of indigenous and local flora. An associated library contains over 6500 books and journals, and the main task of the staff is to classify the native and naturalised flora of the state and to help promote its economic aspects, Western Australia having a million dollar export trade in cut indigenous flora to flower markets around the world. Surrounding the Herbarium building is a 2.7 hectare garden of indigenous plants which was begun in 1970 and is not only decorative, but of value for taxonomic research.

With the development of its first-class herbarium and the Botanic Garden in Kings Park, Western Australia, having lagged behind the other states, has begun to make a major contribution to botany. It is hoped that the absence of trees and plants that grow on laterite soils and heavy clay loam will be rectified if a proposed 60 hectare annexe to the Botanic Garden in Kings Park at Wungong, 40 kilometres south-east of Perth, goes ahead.

Within the Botanic Garden in Kings Park, the existing collection of plants indigenous to the southwest of the state will be extended to include some of the less ornamental species of the region. But it is for the overwhelming beauty of its ornamental flora from Western Australia that the Garden is memorable, not only as a botanical experience but, because the wealth of bird life, the adjacent bushland of the Park, and the views of the city enhance it as a pleasure park.

11

THE AUSTRALIAN NATIONAL BOTANIC GARDENS, CANBERRA

Looking down on the Australian National Botanic Gardens from the telecommunications tower on top of Canberra's Black Mountain, it is hard to distinguish them from the rest of the grey-green eucalytptus-clad reserve below. But drive in the gates at the foot of the mountain, where the gardens occupy 40 hectares of the south-eastern corner, and you will find Australian flora with a diversity of form and colour not seen anywhere else in the country. Here, gathered wild from all parts of the continent, grow some five thousand of Australia's estimated twenty thousand species of indigenous plants, more than in any one spot on earth.

Propagated, 'domesticated' and cultivated, they form a botanic garden that is part systematic botanical collection, part public park, and part Australian bush garden. The gardens developed, as did the National Botanic Garden of South Africa, at Kirstenbosch, Capetown, for the specific study of indigenous flora rather than for economic purposes or for the study of botany in general. This is the only state botanic garden to grow Australian flora exclusively.

A walk here is a total Australian experience. It is the only state botanical garden not on the seaboard, and the air is dry and typical of inland Australia. So is the climate typically continental, with long, cold winters and hot, dry summers and an average temperature ranging from 0.5 degrees C to 11.1 degrees in July and 12.8 degrees to 27.5 degrees in January. (This makes Canberra the mainland's coldest capital city.) There is an average annual rainfall of 640 mm.

The birds here are predominantly native species, approximately 115 of them, and they are typical of inland areas of Australia. The soil is dry and gravelly, and paths of the local Paddy's River gravel curve naturalistically up the mountain, travelling steeply 100 metres upwards with the temperature rising during the climb by as much as 5 degrees C in winter.

You can run the gamut of Australian environments here. There are the swampy areas of sedges and rushes at the bottom of the gardens, the mallee section, the temperate rainforest with its rainbows in the misty air, a rock garden full of dryland plants growing in the crevices of granite boulders; passing through simulated Hawkesbury sandstone country, you reach at the top a nature trail leading into unadulterated Black Mountain bush. Throughout the Gardens some fifty thousand native plants have been introduced into the existing bushland, but the forest effect is not lost.

Some parts of the Gardens are particularly popular. The boardwalk through the temperate rainforest, with its delicate and often very beautiful landscaping of palms and ferns, is a favourite spot; a platform here is one of Canberra's most romantic places for a wedding. For horticulturists, the section near the top of the gardens consisting of flowering native shrubs and groundcovers beneath a forest of indigenous eucalypts, is a spectacular demonstration of the way in which native plants can match the visual effect of understorey azaleas and camellias in northern hemisphere woodland planting. In the spring, the blues, pinks and lemons of native flowers are highlighted with the whiteness of the eucalypt trunks. Of special interest, too, is the rock garden with its granite boulders, waterfall and exciting small plants including the rare white form of the waratah, *Telopea speciosissima*, discovered by naturalist and writer Thistle Harris in a reserve north of Canberra. The collection of *Callitris* species and the Eucalyptus Lawn, featuring many *Eucalyptus* species, are impressive. As you wander through the Gardens you glimpse fine views of the city and Lake Burley Griffin below.

Planting suits the various habitats, or is grouped according to geography, with for instance the native plants of Tasmania together. Taxonomic order, with families of plants grouped together, dictates a great

---

Left: Eucalyptus mannifera ssp. *maculosa.*

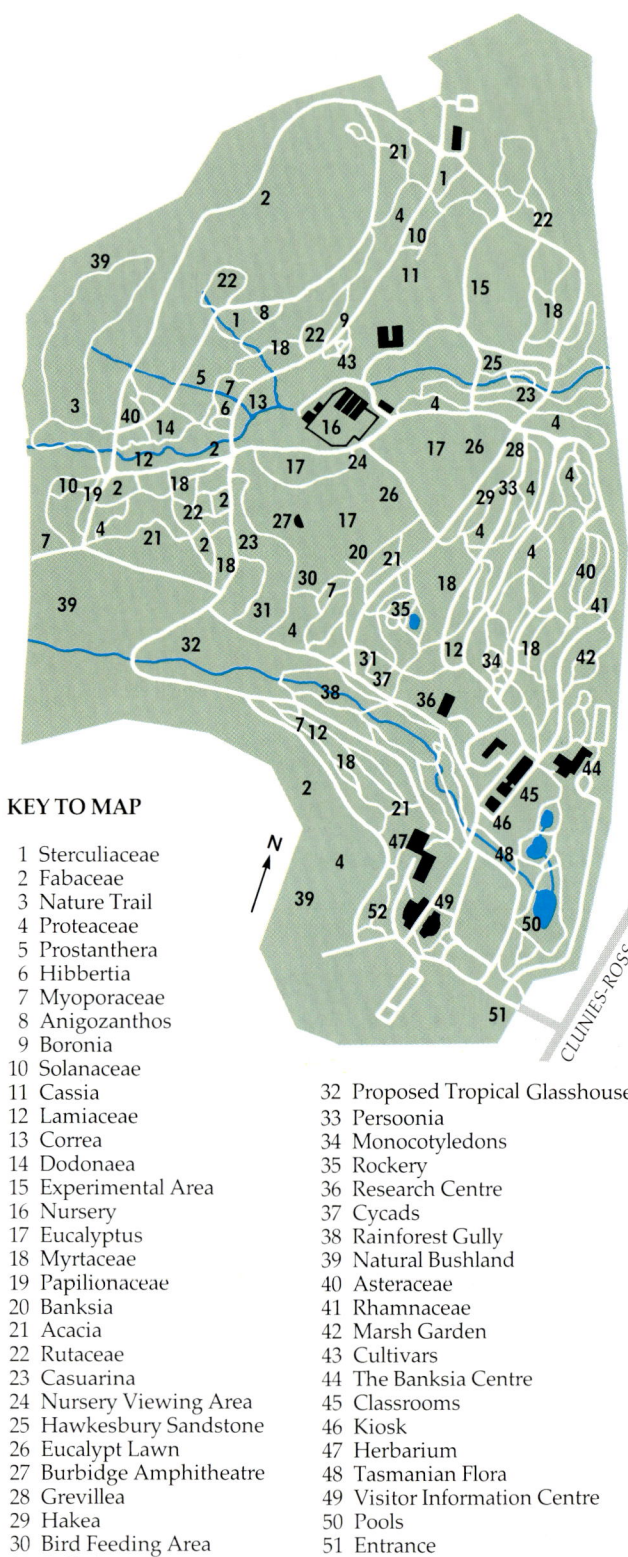

KEY TO MAP

1 Sterculiaceae
2 Fabaceae
3 Nature Trail
4 Proteaceae
5 Prostanthera
6 Hibbertia
7 Myoporaceae
8 Anigozanthos
9 Boronia
10 Solanaceae
11 Cassia
12 Lamiaceae
13 Correa
14 Dodonaea
15 Experimental Area
16 Nursery
17 Eucalyptus
18 Myrtaceae
19 Papilionaceae
20 Banksia
21 Acacia
22 Rutaceae
23 Casuarina
24 Nursery Viewing Area
25 Hawkesbury Sandstone
26 Eucalypt Lawn
27 Burbidge Amphitheatre
28 Grevillea
29 Hakea
30 Bird Feeding Area
31 Callitris

32 Proposed Tropical Glasshouse
33 Persoonia
34 Monocotyledons
35 Rockery
36 Research Centre
37 Cycads
38 Rainforest Gully
39 Natural Bushland
40 Asteraceae
41 Rhamnaceae
42 Marsh Garden
43 Cultivars
44 The Banksia Centre
45 Classrooms
46 Kiosk
47 Herbarium
48 Tasmanian Flora
49 Visitor Information Centre
50 Pools
51 Entrance
52 Mallee Shrubland

The climate of Canberra limits the range of viable native plants. Tropical native plants and many of the species round Sydney are difficult to grow here, but the steepness of the mountain, with its varied terrain, allows microclimates in which plants that would not normally survive the climate can grow. Ninety-five per cent of the planting is from stock gathered by the Gardens' horticulturists and botanists all over Australia. Propagated from seed or cuttings in the Gardens' nurseries, plants are pruned and fertilised as exotics would be, and arranged informally. It is here that the domestic gardener will come for ideas on how to grow native plants on a smaller scale.

Linked to the planting are the dried specimens in the herbarium, which is in the 1974-built Botany Building near the entrance to the Gardens. Here is the most comprehensive and largest collection of dried Australian flora to be seen anywhere, a collection of approximately 160 000 specimens.

According to their founding director, Professor Lindsay Pryor, formerly of the Australian National University in Canberra, these gardens are already highly thought of internationally. They are well sited, and unlike older capital city botanic gardens, they have room to expand. They are also free from pollution and big-city industrial traumas that can affect planting. Their main asset, though, is their exclusive concentration on Australian flora. Eighty-five per cent of Australian native plants belong uniquely to this country. Many species, including three thousand in Western Australia alone, have yet to be named and cultivated, and Australia is one of the few remaining botanical frontiers. To the world's scientists, the insectivorous plants of Western Australia, the Proteaceae of northern Queensland, the boronias and the acacias (particularly the phyllodinous or whole-leaved species) are extraordinarily interesting, and the National Gardens' botanists and horticulturists have a pioneering role to play in studying and reproducing them here.

The observation and registration of new cultivars, derived either from sports or unusual forms of species, or bred by hybridisation, is carried out here, under the auspices of the Australian Cultivar Registration Authority. Each year about thirty new species emanating from botanists' trips into the wild are cultivated at the Gardens and released as new plants to the Australian Nurserymens Association for distribution to the general public.

The decision to grow Australian native flora exclusively at the National Botanic Gardens did not come easily. The original plan for botanical gardens in Canberra, included in Walter Burley Griffin's 1913 scheme for the national capital, was to feature a 'continental arboretum' containing trees from all over the world. A 360 hectare site near the proposed university and lake was put aside on the slopes of Black Mountain. Native plants were not mentioned, although Walter Burley Griffin is known to have been keen on native flora and at one time had a scheme to clothe Black Mountain in natives with

deal of the planting. Educational concepts lie behind other arrangements, as in the section on native flora considered to be noxious weeds in other parts of the world. In some areas plants have been arranged for landscaping effect. Thus, scattered throughout the gardens you come across the same plants in different roles – 50 species and varieties of *Banksia*, 225 of *Acacia*, 92 of *Melaleuca*, and some 360 of *Eucalyptus*. The predominant trees, though, are the local Black Mountain *Eucalyptus mannifera* ssp. *maculosa*, (brittle gum), *E. rossii* (inland scribbly gum) and *E. macrorhynca*, the red stringybark.

Left: *Eucalyptus mannifera* ssp. *maculosa.*
Above: The trunk of a *Eucalyptus rossii,* the inland scribbly gum.

Above right: *Leptospermum scoparium* var. *rotundifolium.*

Right: The blues of *Lechenaultia biloba* and a background of *Dampiera subspicata* beneath a cover of indigenous *Eucalyptus.*

pink and yellow flowers. Nothing, however, was done for decades about the planning or planting of botanical gardens. In the meantime, under the guidance of the Parks and Gardens section of the Department of the Interior, millions upon millions of exotic trees and shrubs, mainly deciduous, filled the city and its surroundings to the extent that visitors exclaimed that the whole of Canberra was one botanical garden. All it lacked was the Latin name tags.

By 1933, with still no official botanic gardens for the national capital, the parliamentary Advisory Council recommended that something should be done. Dr B. T. Dickson, head of the Council for Scientific and Industrial Research, and E. A. Bruce, superintendent of Parks and Gardens, were commissioned to submit a feasibility study. The Dickson report is the cornerstone of the Gardens' history, even though its recommendations were later disregarded in at least one important way. Though it emphasised that there should be an area of the Gar-

dens devoted to the scientific study of Australian flora, there should also be exotics from similar cold-climate countries.

Came the war, and the Gardens report was forgotten until in 1944 Dr Lindsay Pryor, an expert on eucalypts and a forester, was appointed Director of Parks for the ACT and began work on the project. He was impressed by the Dickson report and by the research that had gone into the choice of a site. An area on the eastern side of the mountain, further up than Walter Burley Griffin recommended, was kinder and moister than the rest of Canberra's infamous 'Frost Hollow'. With Canberra's average of twenty-nine severe frosts a year, its low rainfall and fierce, drying westerly winds, there would be a limit anyway to what could be grown in the Gardens.

Dr Pryor eventually chose a site further up the mountain still. The higher up, the less frost-prone it would be. The area contained two gullies that could provide a range of climatic conditions to suit a greater variety of plants than can normally be sustained in Canberra. It had an altitude range of 100 metres between the bottom of the site and the top. Most of the original forest had been ringbarked in the early 1900s, what was there was forest regrowth – so there was no aesthetic objection to clearing some of it.

On 21 September 1945, Dr Pryor got a grant from parliament for £1000 to survey the site, clear stumps and purchase some meteorological equipment. There was no plan or philosophy for the gardens. 'Yet', said Dr Pryor in a recent interview, 'it did seem as though we'd had enough of the English landscaping style of gardening in Canberra and the Gardens should move in another direction.' Dr Pryor heard of the work of Dr Erwin Gauba, an internationally known Viennese botanist who was interned during the war at Loveday camp in South Australia, and who had built up a private herbarium of indigenous plants from the surrounding Mallee district during his internment. Dr Pryor offered Gauba the post of first botanist for the proposed gardens, and he began work on the first National Botanic Gardens herbarium in a private house in Acton. With Lindsay Pryor, he went on field trips collecting local flora round Canberra to be propagated in the government nurseries at Yarralumla. Both scientists thought it was important to have a strong collection of indigenous plants, as, with the spread of cities, the native flora was becoming less accessible to the average citizen.

In 1947 Lindsay Pryor went on a world tour to study botanic gardens. At Göteborg in Sweden he was fascinated by the grouping of plants according to their natural habitat, and met there Dr Bertil Lindquist, the director of the Garden. A plant collector, forester and botanist, he was the creator of the alpine garden of pure species on stony outcrops, and the collection of Far Eastern plants from plant-hunting expeditions in China and Japan. There is also a Japan-Dalan garden, an English-style woodland garden planted exclusively with Japanese native species. A later visit to the Royal Botanic Garden of Edinburgh confirmed Pryor's liking for an ecological style of planting when he saw the renowned rock garden there. He was also fascinated by the Santa Barbara Gardens in California, which grow only plants indigenous to the area. These, he found, had an aesthetic association with the landscape. Fitting other native species into existing vegetation meant lower costs in establishing planting than would result from starting with cleared land. That

the Santa Barbara Gardens were not very big added to their attraction, for the maintenance cost was not high.

Yet even when the Canberra Botanic Gardens as they were then known, were officially begun on the new site on 13 September 1949 during a British Commonwealth Conference on agriculture, held in Canberra, there was still no settled policy to concentrate on native flora. The *Canberra Times* reported at the time that although the emphasis would be on native flora, collections of plants from countries of similar climates would be included. The Prime Minister, Ben Chifley, ceremonially planted an English oak and Sir Edward Salisbury, the director of the Royal Botanic Gardens, Kew, planted a *Eucalyptus mannifera* ssp. *maculosa*. The oak, transplanted later, failed to survive. The brittle gum thrives by the entrance to the Gardens. Planting of the specimens which Dr Pryor and Dr Gauba had been nursing at Yarralumla began. Then, as today, 95 per cent of the plants had originally been gathered from the wild.

On the lower reaches went the collections of acacias, callistemons, grevilleas, hakeas, melaleucas, banksias, and the plants of the Mallee. Dr Pryor chose suitable sites for the healthy growth of plants rather than adhering to a strict landscaping plan. Now, thirty-eight years later, this is the most mature section of the Gardens, with some fine eucalypts and well-grown shrubs, although some short-lived species, such as some acacias, have already died and been replaced.

In 1952, 80 hectares of a farm on the south coast near Jervis Bay, a pocket of New South Wales that belongs to the Commonwealth, was offered to the Botanic Gardens and accepted by Dr Pryor. It had a warm, moist coastal climate with a higher rainfall than Canberra. It would provide a frost-free nursery and ensure that some of the field specimens collected throughout the country would be safely stored and propagated. It contained a range of terrain from heathland to rocky sandstone escarpment with some of the last untouched stands of Hawkesbury sandstone flora, some of the most brilliant in the country, remaining on the coast.

Lindsay Pryor resigned as Director of Parks and Gardens for the ACT in 1958 to become the first professor of botany at the Australian National University, and his place was taken by his deputy, David Shoobridge. A former forester colleague of Dr Pryor, he was related to the Hon. J. M. Shoobridge, who helped build the Hobart conservatory. Shoobridge could see the Gardens eventually becoming as important internationally as Kew. 'In 100 years' time what is planted in the gardens may be unique in the world', he said in an interview.

Field trips by botanists and horticulturists to wild areas with climates similar to Canberra's – places such as inland Western Australia, the Northern Territory and South Australia – increased. The prime targets for collecting were the Crown lands that might one day be cleared for farming. Shoobridge collected plants with his wife Molly, a keen amateur botanist, and the late Dr Betty Phillips, who succeeded Dr Gauba as chief botanist for the Gar-

dens. As a child Dr Phillips had lived next door to Walter Burley Griffin in Melbourne, and she had been chief botanist to the Snowy Mountains Authority. She was responsible for establishing the 'voucher system' of labelling plants. This entails marking incoming plants with tags and numbers as soon as they are collected from the wild, with the colour of the tag denoting whether or not the plant has been identified. Through this system the plant's progress both as a living thing and as a dried herbarium specimen, can be monitored. Dr Phillips's system is believed to be unique in the world.

During the twenty years from 1959 to 1979 the Gardens staff grew from three to sixty. The plant collection expanded from about 4500 to 30 000. Water and irrigation systems developed from one tap and watering cans to a full irrigation system from the town supply and a self-contained internal reservoir to irrigate the high parts. The first nursery, a hand-dug patch at the bottom of the Gardens, grew into the sophisticated propagation and shade houses halfway up. In 1962 the herbarium moved from its wooden house in Acton, which was below the waterline for Lake Burley Griffin (due to be filled in 1963), to a brick house in Downer.

On 27 September 1964 the Gardens were opened for eight hours daily. The official opening by the Prime Minister, John Gorton, was not until September 1970, during the Sixth International Congress of Parks Administration in Canberra. (The name was changed to the National Botanic Gardens in 1978, and to the Australian National Botanic Gardens in 1984.) In 1974, the first permanent building, containing an administration block, laboratory and herbarium, was constructed in the Gardens; permanence had been achieved.

Above left: Acacia howittii, a weeping hedging plant in the Gardens, was once an endangered species.

Above: Callistemon pallidus.

Below: Boronia mollis 'Lorne Pride', a cultivar.

Right: **The double-tailed orchid,** *Diuris punctata* **var.** *albo-violacea,* **an endangered species.**

Far right: Rare endangered species *Darwinia carnea,* the Mogumbar bell, from Western Australia, is cultivated in the Gardens.

The policy of specialising in native flora was established by the early 1960s. When, in 1966, it became obvious that the Gardens would need a full-time curator, a cottage was built in the grounds. After a brief period as curator, Dr Ross Robbins left and the post was filled in 1967 by John Wrigley, an organic chemist and a foundation member of the Society for Growing Australian Plants. He had started the Kuring-gai Wildflower Garden north of Sydney, and was a devotee of the native flora. Wrigley stayed at the Gardens until 1981, developing the path system, helping to design the rock garden and establishing several of the ecological plantings, for example those for the Hawkesbury sandstone area and the Mallee. The Wrigley–Shoobridge team achieved the 'informalising' of the Gardens and its distinctive landscape style became established.

The planting according to habitat continued with the establishment of a temperate rainforest in the lower of the two gullies. This entailed making a new moist microclimate with a complex irrigation system to produce mist at two-minute intervals. Existing trees, from the dry sclerophyll forest of the mountain, were left in place. The sides of the gully were shored up and planted with ferns, and temperate rainforest canopy trees such as *Eucalyptus torelliana, E. robusta, Acacia binervata* and *A. elata* were added. Over a period of some years the old forest trees died and were removed by crane. The new growth flourished. Now staghorns and bird's nest ferns are naturalised in the treetops.

David Shoobridge retired in 1977. He had achieved his aim of seeing the Gardens carried forward by their own momentum and with a new director whose work no longer included the other parks and gardens in Canberra – his successor, Dr Robert Boden. Boden had worked in the Gardens in the early 1960s before joining the Australian National Parks and Wildlife Service, and he has degrees in botanical ecology, conservation and recreation, and thus the qualifications to cope with the traditionally competitive botanical and horticultural sides of botanic gardens management. He inherited gardens with room only to maintain existing species, and organised an environmental study for a 50 hectare site slightly south of the existing gardens for an extension which should be able to sustain a further five thousand species of native plants when finance is available. Dr Boden also established the special garden for the disabled where horticultural therapy programmes have developed.

News and information on the culture of Australian natives are disseminated from the Gardens, using the experience gained by the botanists and horticulturists working there, by means of a series of booklets first published in 1971 and called *Growing Native Plants.*

Since 1974, when it moved from the Royal Botanic Gardens of Melbourne, to the National Botanic Gardens, the Australian Cultivar Registration Authority has been examining cultivars with a view to including them among cultivated varieties permitted to use special names. By 1987 186 cultivars had proved stable enough to be registered, including *Boronia mollis* 'Lorne Pride', discovered by John Wrigley, which flowers prolifically in the upper reaches of the Gardens in November and December. Another important cultivar to be seen in the gardens is the waratah *Telopea* 'Braidwood Brilliant', a cross between the Sydney waratah and the Braidwood species, combining the frost-resistant qualities of the Braidwood species with the brilliant flower of the Sydney waratah, and bred by Dr Boden specifically for the southern climate during his time with the Parks and Gardens section horticulture research unit.

Field trips by botanists and horticulturists continue. A trip taken in 1980 by Gardens staff in conjunction with the Australian National Parks and Wildlife Service, to Kakadu National Park in the Northern Territory, yielded 150 new species native to the park and suitable for landscaping purposes in the Northern Territory. Dr Boden's main concern is, however, like that of Dr Pryor and David Shoobridge, to conserve plants that might otherwise become extinct. Under present conditions, he says,

Right: **Temperate rainforest is an artifically created habitat.**

Below: **Paths of local Paddy's River gravel wind through the Gardens.**

about two hundred species of Australian plants will become extinct within the next decade. It is important to cultivate as many species as possible, and ideally their natural habitats should also be preserved. (Some of the Gardens' most successful landscaping plants were once threatened with extinction, including *Acacia howittii*, used frequently as a screening plant in the Gardens.

Scientists in the Botanic Gardens' laboratories work on ways of simulating natural processes that occur in the wild. Some seeds must be exposed to fire before they can germinate, others must be refrigerated, and others must pass through the digestive system of birds. Scientists have recently managed to simulate growing conditions for the purple double-tailed orchid, one of the world's rare orchids now on the verge of extinction in the grasslands north of Melbourne. Seedlings have been raised to tuber stage. If enough stocks are available they will be distributed to other botanic gardens.

Work at the annexe at Jervis Bay is accelerating and the gardens are now open to the public. The 80 hectares there hold about 1800 plant species, mainly flora typical of the Hawkesbury sandstone country, coastal escarpment heathland and rainforest. They are predominantly a duplicate collection of plants which can only survive in Canberra under glass. Each year about a thousand new plants are planted out, labelled, and protected from kangaroos and possums by wire netting. Irrigation, using water from the freshwater Lake Mackenzie (originally a deep crater in the sand dunes system), is planned.

The landscaping style here is similar to that in Canberra, with clumps of similar-sized shrubs grouped for aesthetic effect in clearings in the bush. *Banksia robur*, *Calothamnus* and *Eriostemon* are predominant. Stands of indigenous trees, the turpentines (*Syncarpia glomulifera*) and blackbutts (*Eucalyptus pilularis*), are conspicuous near the lake. Other areas, particularly around the sandstone escarpment which rings the land falling down to Lake Mackenzie, have been left in a natural bush state. Here you can see the pale pink *Grevillea barklyana*, and the trigger plant, *Stylidium graminifolium*.

Planting is arranged according to habitat. In the heathland at the top of the Gardens, as you approach the car park, the banksias have been controlled by regular firing to allow smaller indigenous heath plants like the Christmas bells, *Blandfordia cunninghamii*, enough light and space to grow in. Round the stone escarpment a donated collection of native orchids is becoming naturalised. A semi-tropical rainforest section has been planted in the creek bed at the centre of the land, using vestiges of indigenous rainforest with additions of Queensland top-storey canopy plants, eucalypts and acacias, and a middle storey of palms and ferns. The best time to visit this Jervis Bay annexe is in late winter and early spring, August and September, while October is the time to visit the main gardens in Canberra, where spring comes later due to the more severe winter.

With native plants, though, there is a wide range of flowering seasons, and the many walks through these gardens show some flowers all through the year. In the Australian National Botanic Gardens, collections are best viewed by the visitor who takes one or all of the designated walks. There are walks showing the trees used by Aborigines for food, medicines and clothing, and, in the nature trail, the planting is linked to interesting features of the environment, for example the anthills.

In 1985 a new two-section building for the Gardens was opened. It contains an administration section and a section catering for the public that includes a public-access herbarium, soon to be completed. Though long-term plans for the Gardens – the large conservatory for tropical plants and 50 hectares in addition to the present site – await the necessary funds, there are short-term projects for improvements to be completed within the next three years. These include the building of an orchid display house for the outstanding collection of native orchids held here, and a herbarium for non-vascular plants.

Already the Gardens live up to their early promise of being an important international reference authority on Australian flora and an attraction for botanists and horticulturists from all over the world. With its botanical excellence, the Australian bush flavour, and the beauty and interest of the vast collection of plant species, the Australian National Botanic Gardens have earned a place as one of the most distinguished institutions of the nation's capital, and they are now part of the Department of Arts, Heritage and the Environment.

Left: *Eriostemon australasius* at the Jervis Bay annexe.

Right: Tree ferns on the creek bed at the Jervis Bay annexe.

Below: The freshwater Lake Mackenzie at the Jervis Bay annexe.

Overleaf: Rock escarpments at the Jervis Bay annexe, a southern extremity of Hawkesbury sandstone country.

BIBLIOGRAPHY

Adams, Brian. *The Flowering of the Pacific*. Collins, Sydney, 1986.

Aitken, Richard. The Australian Garden History Society Conference, Ballarat, 1984.

Baker, K. C., curator. *The Garden Story: an historical review of a century of development at the Rockhampton Botanic Gardens*. Rockhampton City Council, 1969.

Bauer, June. *Some Other Eden: A History of the Darwin Botanic Garden*. North Australia Research Bureau, ANU, Canberra, 1980.

Blombery, Alec and Rodd, Tony. *Palms*. Angus and Robertson, Sydney, 1982.

Bloomfield, Paul. *Edward Gibbon Wakefield*. Longmans, Green and Co Ltd, 1961.

Blunt, Wilfrid. *In for a Penny: A Prospect of Kew Gardens*. Hamish Hamilton, Tryon Gallery, London, 1978.

Canberra Botanic Gardens. Canberra Department of the Interior. 1970.

Capon, Joanna. *Report on the garden of Old Government House*. Sydney, 1986.

Churches, David. The Cultural Significance of the Royal Botanic Gardens, Sydney, including its structures. Master's thesis, University of NSW, 1987.

Erickson, Rica. *The Drummonds of Hawthorden*. Lamb Peterson, Osborne Park, WA, 1969.

Everard, Barbara W. and Morley, Brian D. *Wildflowers of the World*. Ebury Press and Michael Joseph, London, 1970.

Gibson-Wilde, Dorothy M. Gateway to a Golden Land. History Department, James Cook University, Townsville, North Queensland, 1984.

Gibson-Wilde, Dorothy M. Notes on Eugene Fitzallan (sic), (unpublished manuscript).

Gilbert, Lionel. Botanical Investigation of New South Wales 1811–1880. Unpublished Ph.D. thesis.

Gilbert, Lionel. 'Plants politics and personalities in Colonial New South Wales', essay in D. J. Carr and S. G. M. Carr, *People and Plants in Australia*. Academic Press, Sydney, 1981.

Gilbert, Lionel. *The Royal Botanic Gardens, Sydney: A History 1816–1985*. Oxford University Press, Melbourne, 1986.

Herbert, D. A. 'The Brisbane Botanic Garden', The C. T. White Memorial Lecture reprinted from *The Queensland Naturalist*, Vol XIV, No 4.

Hughes, Verna Joyce. The Custody of King's Park 1929–1931. Thesis, 1978.

Hurburgh, Marcus. *The Royal Tasmanian Botanical Gardens 1818–1986: A History in stone, soil and superintendents*. Shearwater Press, Sandy Bay, Tasmania, 1986.

Hyams, Edward. *A History of Gardens and Gardening*. Preager Press, New York, 1971.

Hyams, Edward and McQuitty, William. *Great Botanical Gardens of the World*. Thomas Nelson, London, 1969.

Jacobs, Wendy, et al. *Ballarat: A Guide to Buildings and Areas 1851–1940*. Jacobs Lewis Vines architects and conservation planners, South Yarra, Victoria, 1981.

Jellicoe, Sir Geoffrey and Jellicoe, Susan, et al. *The Oxford Companion to Gardens*. Oxford University Press, Oxford, 1986.

Jervis, James. *The Cradle City of Australia*. Parramatta, 1961.

Jones, George. *Growing Together: A Gardening History of Geelong Extending to Colac and Camperdown*. George Jones, Belmont, Victoria, 1984.

Kynaston, Edward. *A Man on Edge: A Life of Baron Sir Ferdinand von Mueller*. Penguin Books, Melbourne, 1981.

Lamshed, Max. *The People's Garden: A Centenary history of the Adelaide Botanic Garden 1855–1955*, The Botanic Garden, Adelaide.

Law-Smith, Joan. *The Royal Botanic Gardens, Melbourne*. Maud Gibson Trust in association with the Royal Botanic Gardens Melbourne, Melbourne, 1984.

Lemmon, Kenneth. *The Golden Age of Plant Hunters*. Phoenix House, London, 1968.

Lord, E. E. *Shrubs and Trees for Australian Gardens*. Melbourne, 1970.

Lumley, Peter and Spencer, Roger. 'Trees in Botanic Gardens: Royal Botanic Gardens, Melbourne', *Arboricultural Journal*, Vol 6, 1982.

Macdonald, Lorna. *Rockhampton: A History of City and District*. University of Queensland Press, St Lucia, Queensland, 1981.

Macoboy, Stirling and Blombery, Alec. *Australian Complete Book of Flowers*. Paul Hamlyn, Sydney, 1975.

Macoboy, Stirling. *What Tree is That?* Lansdowne Press, Sydney, 1979.

Morrison, Crosbie. *Melbourne's Garden*. Melbourne University Press, Melbourne, 1955.

Moyal, Anne. *A Bright and Savage Land*. Collins, Sydney, 1986.

North, Marianne. *A Vision of Eden*. Webb and Bower, Exeter, 1980.

Pescott, R. T. M. *The Royal Botanic Gardens Melbourne, A History from 1845–1970*. Oxford University Press, Melbourne, 1982.

Pescott, R. T. M. *W R Guilfoyle 1840–1912: Master of Landscaping*. Oxford University Press, Melbourne, 1974.

Prest, John. *The Garden of Eden: The Botanic Garden and the Recreation of Paradise*. Yale University Press, New Haven and London, 1981.

Proudfoot, Helen. *Old Government House: The Building and its Landscape*. State Planning Authority of NSW in association with Angus & Robertson, Sydney, 1971.

Stearn, W. T. 'The Introduction of Plants into the Gardens of Western Europe during 2000 years'. Supplement to the *Australian Garden Journal*, April, 1984.

Tanner, Howard and Begg, Jane. *The Great Gardens of Australia*. Macmillan, South Yarra, 1976.

Tench, Captain Watkin. 'A Complete Account of a settlement of Port Jackson in New South Wales'. London, 1793.

Trollope, Anthony. *Australia and New Zealand*. George Robertson, Melbourne, 1873.

Wakefield, Edward Gibbon. *A View of the Art of Colonisation*. 1860.

Walch's Tasmanian Guide Book: Handbook of information for all parts of the colony. Hobart, Walch, 1871.

Watts, Peter. *Historic Gardens of Victoria: A Reconnaissance*. Oxford University Press, Melbourne, 1983.

 # INDEX